NEW YORK TIMES B[...]

FRANCES E. JE[...]

with AMY ELLIS NUTT

THE TEENAGE BRAIN

A NEUROSCIENTIST'S SURVIVAL GUIDE TO RAISING ADOLESCENTS AND YOUNG ADULTS

"Frances Jensen, a neuroscientist and single mother of two boys . . . delved into the emerging science of the adolescent brain [and] came out with provocative new insights for parents, educators, public policymakers, and teens themselves." —*Washington Post*

THE TEENAGE BRAIN

A Neuroscientist's Survival Guide to Raising Adolescents and Young Adults

Frances E. Jensen, MD

with Amy Ellis Nutt

HARPER

NEW YORK • LONDON • TORONTO • SYDNEY

HARPER

A hardcover edition of this book was published in 2015 by HarperCollins Publishers.

This book is designed to give information on various medical conditions, treatments, and procedures for your personal knowledge and to help you be a more informed consumer of medical and health services. It is not intended to be complete or exhaustive, nor is it a substitute for the advice of your physician. You should seek medical care promptly for any specific medical condition or problem you may have. All efforts have been made to ensure the accuracy of the information contained in this book as of the date published. The authors and the publisher expressly disclaim responsibility for any adverse effects arising from the use or application of the information contained herein.

HarperCollins books may be purchased for educational, business, or sales promotional use. For information, please e-mail the Special Markets Department at SPsales@harpercollins.com.

FIRST HARPER PAPERBACK EDITION PUBLISHED 2016.

Designed by JoAnne Metsch

Library of Congress Cataloging-in-Publication Data has been applied for.

ISBN 978-0-06-206785-2 (pbk.)

17 18 19 20 OV/RRD 10 9 8

PRAISE FOR

The Teenage Brain

"It's charming to see good science translate directly into good parenting."
—*New York Times Book Review*

"Frances Jensen, a neuroscientist and single mother of two boys . . . delved into the emerging science of the adolescent brain [and] came out with provocative new insights for parents, educators, public policymakers, and teens themselves."
—*Washington Post*

"[Jensen] offers a parenting guide laced with the latest MRI studies. . . . Packed with charts and statistics."
—Elizabeth Kolbert, *The New Yorker*

"This well-written, accessible work surveys recent research into the adolescent brain, a subject relatively unexplored until just this past decade. . . . Speaking as one parent to another, she offers support and a way for parents to understand and relate to their own soon-to-be-adult offspring."
—*Publishers Weekly*

"Meticulously detailed and reported, the studies offer proof that it's not just parents who think their teenagers don't quite have it all together."
—*Kirkus Reviews*

"A captivating chapter, 'The Digital Invasion of the Teenage Brain,' calls attention to computer craving and adolescent addiction to the Internet. . . . [A] sensible, scientific, and stimulating book."
—*Booklist*

"A valuable resource for parents, youth workers, educators, and anyone involved with teens in any way. The book is engaging, understandable, and extremely informative."

—*New York Journal of Books*

"My favorite quote from this marvelous book: 'The truth of the matter is . . . adolescents are not an alien species, just a misunderstood one.' Dr. Jensen uses her considerable expertise as a neuroscientist and a mother to explain the recent explosion of adolescent brain research and how this research can help us better understand and help young people. This book also highlights biologically inherent opportunities to enhance the health and well-being of young people during the second decade of life . . . opportunities we should not be missing."

—Carol A. Ford, MD, President, Society for Adolescent Health and Medicine; Professor of Pediatrics, University of Pennsylvania; and Chief, Division of Adolescent Medicine at the Children's Hospital of Philadelphia

"In *The Teenage Brain*, neurologist Frances Jensen has brilliantly translated academic science and clinical studies into easily understandable chapters to highlight the many changes in connections and plasticity of the brain. The book is a 'must read' for parents, teachers, school nurses, and many others who live with or interact with teens."

—S. Jean Emans, MD, Chief, Division of Adolescent/Young Adult Medicine, Boston Children's Hospital; Professor of Pediatrics, Harvard Medical School

This book is dedicated to my two sons, Andrew and Will. Watching them grow into young men as they emerged through their teen years has been the joy of my life, and shepherding them through this time was probably the most important job of my life. Together we went on a journey, and as much as I taught them, they taught me. The product is this book, and I hope that it informs not only those people helping to raise adolescents, but also the teenagers themselves.

When I was a boy of fourteen, my father was so ignorant I could hardly stand to have the old man around. But when I got to be twenty-one, I was astonished by how much he'd learned in seven years.

—MARK TWAIN

I would that there were no age between sixteen and three-and-twenty, or that youth would sleep out the rest, for there is nothing in the between but getting wenches with child, wronging the ancientry, stealing, fighting . . .

—*The Winter's Tale*,
WILLIAM SHAKESPEARE

Contents

Illustrations

THE TEENAGE BRAIN

Introduction
Being Teen

What was he thinking?

My beautiful, auburn-haired son had just returned home from a friend's house with his hair dyed jet-black. Despite my inward panic, I said nothing.

"I want to get red streaks in it," he told me nonchalantly.

I was gob-smacked. *Is this really my child!?* I'd begun to ask the question often during my fifteen-year-old son Andrew's sophomore year at a private high school in Massachusetts, all the while trying to be empathetic. I was a divorced working mother of two teenage sons, putting in long hours as a clinician and professor at Boston Children's Hospital and Harvard Medical School. So if I sometimes felt guilty about the time I spent away from my boys, I also was determined to be the best mother I could be. After all, I was a faculty member in a pediatric neurology department and actively researching brain development. Kids' brains were my business.

But my sweet-natured firstborn son had suddenly become unfamiliar, unpredictable, and bent on being different. He had just transferred from a very conventional middle school that went through ninth grade, where jackets and ties were the norm, to a very progressive high school. Upon arriving, he took full advantage of the new environment, and part of that was to dress in what you might describe as an "alternative" style. Let's face it, his best friend had spiky blue hair. Need I say more?

I took a deep breath and tried to calm myself. Getting mad at him, I knew, wouldn't do either of us any good and probably would only alienate him further. At least he felt comfortable enough to tell me about something he wanted to do before he actually did it. This was an opportunity, I realized, and I quickly seized it.

Instead of damaging your hair with some cheap, over-the-counter dye, what if I take you to my hair guy for the red streaks? I asked him. Since I also was going to pay for it, Andrew happily agreed. My hair stylist, who was a sort of punk rocker himself, got totally into the task. He did a great job, actually—so good that Andrew's girlfriend at the time was inspired to color her hair in exactly the same black-and-red motif. She attempted this herself, and needless to say had different results.

Thinking back to those days, I realize so much of what I thought I knew about my son during this turbulent time of his life seemed turned on its head. (Was that a compost pile in the middle of his bedroom, or laundry?) Andrew seemed trapped somewhere between childhood and adulthood, still in the grip of confusing emotions and impulsive behavior, but physically and intellectually more man than boy. He was experimenting with his identity, and the most basic element of his identity was his appearance. As his mother and a neurologist, I thought I knew everything there was to know about what was going on inside my teenager's head. Clearly I did not. I certainly didn't know what was going on outside his head either! So as a mother and a scientist, I decided I needed to—I *had* to—find out.

Professionally, I was primarily studying the brains of babies at that time and running a research lab largely devoted to epilepsy and brain development. I was also doing translational neuroscience, which means, simply, trying to create new treatments for brain disorders. Suddenly, however, I had a new scientific experiment and project: my sons. My younger son, Will, was just two years Andrew's junior. What would I be in for when Will reached the same age as his older brother? There was so much I didn't get. I had watched Andrew, almost overnight, morph into a different being, yet I knew, deep inside, he was still the same wonderful, kind, bright kid he'd always been. So what happened? To figure it out I decided to delve into the world of research on this somewhat foreign species in my household called the teenager, and use that knowledge to help me and my sons navigate their way more smoothly into adulthood.

The teen brain has been a relatively neglected area of study until only the past decade. Most research dollars in neurology and neuropsychology are spent on infant and child development—from learning disabilities to early enrichment therapy—or, at the other end of the spectrum, on diseases of the elderly brain, especially Alzheimer's. Up until just a few years ago, the neuroscience of the adolescent brain was underfunded, underresearched, and obviously not well understood. Scientists believed—incorrectly, as it turned out—that brain growth was pretty much complete by the time a child started kindergarten; this is why for the past two decades parents of infants and toddlers, trying to get a jump on their children's education, have inundated their kids with learning tools and accessories like Baby Einstein DVDs and Baby Mozart Discovery Kits. But the adolescent brain? Most people thought it was pretty much like an adult's, only with fewer miles on it.

The problem with this assumption is that it was wrong. Very wrong. There are other misconceptions and myths about the teenage brain and teenage behavior that are now so ingrained they are accepted societal beliefs: teens are impulsive and emotional because of surging hormones; teens are rebellious and oppositional because they want to be difficult and different; and if teenagers occasionally drink too much alcohol without their parents' consent, well, their brains are resilient, so they'll certainly rebound without suffering any permanent effects. Another assumption is that the die is cast at puberty: whatever your IQ or apparent talents may be (a math or science type versus a language arts type), you stay that way for the rest of your life.

Again, all wrong. The teen brain is at a very special point in development. As this book will reveal, I learned that there are unique vulnerabilities of this age window, but there is also the ability to harness exceptional strengths that fade as we enter into adulthood.

The more I studied the emerging scientific literature on adolescents, the more I understood how mistaken it was to look at the teenage brain through the prism of adult neurobiology. Functioning, wiring, capacity—all are different in adolescents, I learned. I was also aware that this new science of the teenage brain wasn't reaching most parents, or at least wasn't reaching parents who don't have a background in neuroscience as I did. And this was just the audience who needed to know about this new science of the adolescent brain: parents and guardians and educators who are just as perplexed, frustrated, and maddened by the teenagers in their care as I was.

When my younger son, Will, was sixteen, he passed his driver's test. He'd rarely given me cause to worry, but that changed early one morning. A few weeks after getting his license, he had started to drive himself to school in our 1994 Dodge Intrepid—a big, old, safe car. All seemed to go well. As usual, Will left around 7:30, as school started at 7:55. Off he went. Just as I was walking out the door to go to my job, at about 7:45, I got a call from Will: "Mom, I'm okay, but the car is totaled." Well, first, I was thankful he had the presence of mind to lead off with telling me he was okay, but I had visions of his car wrapped around a tree. I said, "I'm on my way," and jumped in my car. As I was approaching the school entrance, I saw the flashing lights of the police cars. What had he done? Well, simply put, he had decided that he could squeeze a left turn into the school entrance in the path of rapidly moving traffic going in the opposite direction. This might have worked if there had been another mother like me driving in the opposite direction who would have shaken her head and slammed on her brakes. But in Will's case that morning, it was a twenty-three-year-old guy, a construction worker in a Ford F-150 on his way to work. He was no more in the mood to give the right-of-way than Will had been to wait to cross the road. So—the accident happened. It was good to know that 1994 airbags still worked in 2006.

There was Will standing by his completely trashed car at the very entrance of his school, looking sheepish as basically the entire school drove by him as students and staff arrived for the day. What a lesson for Will. I recognized that immediately—and was so thankful that he and the other driver had emerged unscathed from this battle of wills as to who had the right-of-way.

What was he thinking? I asked myself, almost reflexively.

Then: *Oh, no, here we go again.*

This time, however, I quickly calmed myself. I knew a lot more now. I knew Will's brain, like Andrew's, like every other teenager's, was a work in progress. He clearly was no longer a child, and yet his brain was still developing, changing, even growing. I hadn't recognized that until Andrew made me sit up and take stock of what I knew about the pediatric brain, that it's not so much what *is* happening inside the head of an adolescent as what is *not*.

The teenage brain is a wondrous organ, capable of titanic stimulation and stunning feats of learning, as you will learn in this book. Granville Stanley Hall, the founder of the child study movement, wrote in 1904 about the exuberance of adolescence:

> These years are the best decade of life. No age is so responsive to all the best and wisest adult endeavor. In no psychic soil, too, does seed, bad as well as good, strike such deep root, grow so rankly or bear fruit so quickly or so surely.

Hall said optimistically of adolescence that it was "the birthday of the imagination," but he also knew this age of exhilaration has dangers, including impulsivity, risk-taking, mood swings, lack of insight, and poor judgment. What he couldn't possibly have anticipated back then is the breathtaking range of dangers teens would be exposed to today through social media and the Internet. How many times have I heard from friends, colleagues, even strangers who have reached out to me after hearing me speak, about the crazy things their teenage kids or their friends just did? The daughter who "stole" her father's motorcycle and crashed it into a curb. The kids who went "planking"—lying facedown, like a board, on any and every surface (including balcony railings), and then taking photos of one another doing it. Or worse: "vodka eyeballing," pouring liquor directly into the eye to get an immediate high, or, scared about passing a drug test for a weekend job, ingesting watered-down bleach, thinking it would "clean" their urine of the pot they had smoked the night before.

Children's brains continue to be molded by their environment, physiologically, well past their midtwenties. So in addition to being a time of great promise, adolescence is also a time of unique hazards. Every day, as I will show you, scientists are uncovering ways in which the adolescent brain works and responds to the world differently from the brain of either a child or an adult. And the way that the adolescent brain responds to the world has a lot to do with the impulsive, irrational, and wrongheaded decisions teens seem to make so frequently.

Part of the problem in truly understanding our teenagers lies with us, the adults. Too often we send them mixed messages. We assume that when our kid begins to physically look like an adult—she develops breasts; he has facial hair—then our teenager should act like, and be treated as, an adult with all the adult responsibilities we assign to our own peers. Teenagers can join the military and go to war, marry without the consent of their parents, and in some places hold political office. In recent years, at least seven teens have been elected mayors of small towns in New York, Pennsylvania, Iowa, Michigan, and Oregon. Certainly the law often treats teens as adults, especially when those teens are accused of violent crimes and then tried in adult criminal courts. But in myriad ways we also treat our teens like children, or at least like less than fully competent adults.

How do we make sense of our own conflicting messages? *Can* we make sense?

For the past few years I've given talks all over the country—to parents, teens, doctors, researchers, and psychotherapists—explaining the risks and rewards that pertain to the new science of the adolescent brain. This book was prompted by the tremendous, even overwhelming, number of responses I have received from parents and educators (and sometimes even teens) who heard me speak. All of them wanted to share their own stories, ask questions, and try to understand how to help their kids—and, in the process, themselves—navigate this thrilling but perplexing stage of life.

The truth of the matter is, I learned from my own sons that adolescents are not, in fact, an alien species, but just a misunderstood one. Yes, they are different, but there are important physiological and neurological reasons for those differences. In this book I will explain how the teen brain offers *major advantages* on the one hand but unperceived and often *unacknowledged vulnerabilities* on the other. I am hoping you will use this as a handbook, a kind of user's manual or survival guide to the care and feeding of the teenage brain. Ultimately, I want to do more than help adults better understand their teenagers. I want to offer practical advice so that parents can help their teenagers, too. Adolescents aren't the only ones who must navigate this exciting but treacherous period of life. Parents, guardians, and educators must, too. I have—twice. It is humbling, exhilarating, confusing, all at the same time. As parents, we brace ourselves for what will be quite a roller-coaster ride, but in the vast majority of cases the ride slows down, evens out, and gives one a lot of stories to tell afterward!

Nearly a decade ago, when it became clear to me that being a parent of teenagers was nothing like taking care of overgrown children, I said, Okay, let's work on it together. I stayed in my sons' faces. I remember one time, when Andrew was still a sophomore in high school, the inevitable point arrived when exams were just around the corner and he was still paying more attention to sports and parties than books and homework. Because I'm a scientist, I know learning is cumulative—everything new is based on something you just learned, so you have to hang in there, you have to stay on top of it. So I got a pad of paper and I went through each chapter of Andrew's textbooks, and on one side of the paper I picked out a problem for him to solve and on the other side, folded, was the answer. All he needed was a model, a template, a structure. It was a turning point for him *and* me. He realized he actually had to do the work—sit down and do it—in order to learn. He also realized working on his bed, with everything spread out around him, wasn't helping. He needed more structure, so he sat himself at his desk, with a pencil sharpener and a piece of paper in front of him, and he learned how to impose order on himself. He needed the external cues. I could plan and he couldn't at that point. Having a structured environment helped him learn, and eventually he got really good at it, sitting in his chair at his desk for hours. I know because I'd check in on him. I also knew this was a good example of place-dependent learning. Scientists have shown that the best way to remember what you've learned is to return to the place where you learned it. For Andrew, that was his desk in his bedroom. As I will explain later, teenagers are "jacked up" on learning—their brains are primed for knowledge—so where and how they learn is important, and setting up a place where homework is done is something any parent can help teens do. And because homework is one of the main things kids do at home, you can stay involved with your teenagers even if you don't happen to have an MD or PhD in the subject or subjects they've neglected for months. You can offer to proofread assignments, spell-check their essays, or simply make sure they are sitting in a comfortable desk chair. While you might not have the right hair guy to get red streaks, the point is that you can at least spring for a home hair dye when they want to transform themselves on the outside. Let them experiment with these more harmless things rather than have them rebel and get into much more serious trouble. *Try not to focus on winning the battles when you should be winning the war*—the endgame is to help get them through the necessary experimentation that they instinctively need without any long-term adverse effects. The teen years are a great time to test where a kid's strengths are, and to even out weaknesses that need attention.

What you don't want to do is ridicule, or be judgmental or disapproving or dismissive. Instead, you have to get inside your kid's head. Kids all have something they're struggling with that you can try to help. They can be all over the place: forgetting to bring their books home, crumpling important notes in the bottom of their backpacks, misconstruing homework assignments. Sometimes—or most of the time—they are just not organized, not paying attention to the details of what's going on around them, and so expecting them to figure out how to do their homework can actually be expecting too much. Your teenagers won't always accept your advice, but you can't give it unless you're there, unless you're trying to understand how they're learning. Know that they are just as puzzled by their unpredictable behavior and the uneven tool kit they call their brain. They just aren't at a point where they will tell you this. Pride and image are big for teens, and they are not able to look into themselves and be self-critical.

That's what this book is all about—knowing where their limits are and what you can do to support them. So that you won't get angry or confused at your teens or simply throw your hands up in surrender, I want to help you understand what makes them so infuriating. Much of what is in this book will surprise you—surprise you because you probably thought teenagers' recalcitrant behavior was something they could, or at least should, be able to control; that their insensitivity or anger or distracted attitude was entirely conscious; and that their refusal to hear what you suggest or request or demand they do was entirely willful. Again, none of these things are true.

The journey I will take you on in this book will actually shock you at times, but by the end of the journey I promise you will gain insight into what makes your teens tick because you'll have a much better understanding of how their brains work. I make an effort in this book to reveal, wherever possible, the *real* data from *real* science journal articles. There is much data out there that has not been "translated" for the public. Even more important, the teen generation is one that holds information in great esteem. So when you talk to teens, you owe it to them to have actual *data*. I inserted as many figures into this book as I could where the actual science is shown, and I point out where it applies to our knowledge of the strengths and weaknesses of being a teen. There are lots of myths about teenagers out there that need to be debunked: this book is an attempt to chip away at those myths and explore the new science that is available to inform us.

For this book to be truly effective, however, you must remember a simple rule: First, count to ten. It became a kind of mantra for me when I was raising my sons. But it means more than just taking a deep breath. Let me explain. In leadership courses I've taken for my professional career, one theme that is always emphasized is the Boy Scouts' motto, "Be prepared." I learned in these seminars that the average time an American businessperson spends preparing for a meeting is about two minutes. We probably spend more time just scheduling those meetings than actually thinking about what we're going to say or do in them. I don't mean the big presentations. I mean the one-on-one encounters, which we too often step into cavalierly without taking much time for reflection beforehand. When I heard this statistic, initially it shocked me, but then I thought about my own professional world, where I am the head of a large university neurology department and have my own lab with many graduate and postgraduate students, and I realized, Yep, that's pretty much what does happen. Not a lot of time is devoted to planning or "rehearsing" for all those one-on-one encounters with colleagues and staff, and yet it's these more personal, more direct interactions that often play a pivotal role in the success of an organization. Similarly, the impression you give others in these encounters can affect the direction your career takes; this is why it's so important to plan ahead, at least for more than just a few minutes, and think about how the other person will react during one of these meetings. In your mind, go through what you want to say, step by step, and imagine the range of responses. Now imagine that the other person is your teenage son or daughter. Being prepared for both positive and negative responses will help guide you as you consider your options about what to say or do next. If you appear hotheaded or mentally disorganized, you lose credibility, whether it's with a colleague, an employee, or your teenager.

For parents or teachers, or anyone who has a caring relationship with a teenager, reading this book will arm you with facts—and with fortitude. Changing the behavior of your teen is partly up to you, so *you* have to come up with a plan of action and a style of action that fits your household and your kids, as well as your needs and wants. Remember, you are the adult, and if your child is under eighteen, you also are legally responsible for that "child." Certainly the courts will hold you accountable for your child and, by extension, for the environment you provide for him or her. So take the lead, take control, and try to think for your teenage sons and daughters until their own brains are ready to take over the job. The most important part of the human brain—the place where actions are weighed, situations judged, and decisions made—is right behind the forehead, in the frontal lobes. This is the last part of the brain to develop, and that is why you need to be your teens' frontal lobes until their brains are fully wired and hooked up and ready to go on their own.

But the most important advice I want to give you is to stay involved. As the mother of two sons I adore, I couldn't physically maneuver them into doing what I wanted them to do when they were teenagers, not in the way I could when they were small children. Eventually they were simply too big to just pick up and put down where I wanted them to be. We lose physical control as children leave childhood. Our best tool as they enter and move through their adolescent years is our ability to advise and explain, and also to be good role models. If there's anything I've learned with my boys, it's that no matter how distracted or disorganized they seemed to be, no matter how many assignments they forgot to bring home from school, they were watching me, taking the measure of their mom as well as all the other adults around them. I will talk much more about this later in the book, but just so you know, it all turned out okay in my life and the lives of my sons. Here's the bottom line on my two "former teenagers": Andrew graduated from Wesleyan University with a combined MA-BA degree in quantum physics in May 2011 and is now in a joint MD-PhD program. Will graduated from Harvard in 2013 and landed a business-consulting job in New York City. So, yes, you can survive your teenagers' adolescence. And so can they. And you will all have a lot of stories to tell after it's all over.

1
Entering the Teen Years

In July 2010 I received an e-mail from the frustrated mother of a nineteen-year-old who had just finished his freshman year of college. The mother had heard me give a talk to parents and teachers in Concord, Massachusetts, about the teenage brain, and her e-mail expressed a wide range of emotions, from sadness to confusion to anger, about the boy, whose behavior had suddenly become downright "weird."

"My son gets angry easily," she wrote. "He puts a wall around him so he would not talk. He stays up all night and sleeps all day. He stops doing things he used to enjoy. . . . He was once charming, intelligent, outgoing. These days, good mood is rare. I thought I did all that hard work to raise him, to send him to a very good college, and it all ended up like this."

The woman ended her e-mail with a simple question: "How do I help him?"

Letters and e-mails and calls like these are what prompted me to write this book. Nine months after that mother asked how she could help her son, I received a similar e-mail, this time from the mother of an eighteen-year-old girl. Her daughter, who had once seemed so levelheaded, she wrote, had let her grades slip in high school. She became defiant, ran away from home, and was hospitalized for depression. "This year has been difficult for us," the mother wrote. "Sometimes it seems as if she has been replaced by an alien. It is because of the behavior and the things that she says. She is a completely different person."

I knew how these women felt. At one time, I felt helpless, too. Because I was newly divorced as my older son, Andrew, entered adolescence, I was painfully aware that my children's future, as well as their present, was largely up to me. There was no pulling my hair out and saying, "Go talk to your father about it!" When you're a single parent, the buck stops with you. As parents, we want to open a few doors for our kids—that's all, really. To gently nudge them in the right direction. During their childhood, everything seems to go pretty much by plan. Our kids learn what's appropriate and what isn't, when to go to bed and when to get up in the morning, what not to touch, where not to go. They learn the importance of school, of being polite to their elders, and when they are physically hurt or emotionally wounded, they come to us seeking solace.

So what happens when they reach fourteen, fifteen, or sixteen years old? How is it that the cute, even-tempered, happy, and well-behaved child you've known for more than a decade is suddenly someone you don't know at all?

These are a few of things I say to parents right off the bat: The sense of whiplash you are feeling is not unusual. Your children are changing, and also trying to figure themselves out; their brains and bodies are undergoing extensive reorganization; and their apparent recklessness, rudeness, and cluelessness are not totally their fault! Almost all of this is neurologically, psychologically, and physiologically explainable. As a parent or educator, you need to remind yourself of this daily, often hourly!

Adolescence is a minefield, for sure. It is also a relatively recent "discovery." The idea of adolescence as a general period of human development has been around for aeons, but as a discrete period between childhood and adulthood it can be traced back only as far as the middle of the twentieth century. In fact the word "teenager," as a way of describing this distinct stage between the ages of thirteen and nineteen, first appeared in print, and only in passing, in a magazine article in April 1941.

Mostly for economic reasons, children were considered mini-adults well into the nineteenth century. They were needed to sow the fields, milk the cows, and split the firewood. By the time of the American Revolution half the population of the new colonies was under the age of sixteen. If a girl was still single at eighteen, she was considered virtually unmarriageable. Well into the early twentieth century, children over the age of ten, and often children much younger, were capable of most kinds of work, either on the farm or later in city factories—even if they needed boxes to stand on. By 1900, with the Industrial Revolution in full swing, more than two million children were employed in the United States.

Two things in the decades spanning the middle of the twentieth century—the Great Depression and the rise of high schools—not only changed attitudes about the meaning of childhood but also helped to usher in the era of the teenager. With the onset of the Depression after the stock market crash of 1929, child laborers were the first to lose their jobs. The only other place for them was school, which is why by the end of the 1930s, and for the first time in the history of American education, most fourteen- to seventeen-year-olds were enrolled in high school. Even today, according to a 2003 survey by the National Opinion Research Center, Americans regard finishing high school as the number one hallmark of adulthood. (In much of the United Kingdom a teenager is treated as an adult even if he or she does not finish high school, and in England, Scotland, and Wales it is legal not only to leave school at age sixteen but to leave home and live independently as well.) In the 1940s and '50s, American youth, most of whom were not responsible for the economic survival of their families, certainly did not seem like adults—at least not until they graduated from high school. They generally lived at home and were dependent on their parents, and as more and more children found themselves going to school beyond the eighth grade, they became a kind of class unto themselves. They looked different from adults, dressed differently, had different interests, even a different vocabulary. In short, they were a new culture. As one anonymous writer said at the time, "Young people became teenagers because we had nothing better for them to do."

One man foresaw it all more than one hundred years ago. The American psychologist Granville Stanley Hall never used the word "teenager" in his groundbreaking 1904 book about youth culture, but it was clear from the title of his fourteen-hundred-page tome—*Adolescence: Its Psychology and Its Relations to Physiology, Anthropology, Sociology, Sex, Crime, Religion and Education*—that he regarded the time between childhood and adulthood as a discrete developmental stage. To Hall, who was the first American to earn a PhD in psychology, from Harvard University, and the first president of the American Psychological Association, adolescence was a peculiar time of life, a distinct and separate stage qualitatively different from either childhood or adulthood. Adulthood, he said, was akin to the fully evolved man of reason; childhood a time of savagery; and adolescence a period of wild exuberance, which he described as primitive, or "neo-atavistic," and therefore only slightly more controlled than the absolute anarchy of childhood.

Hall's suggestion to parents and educators: Adolescents shouldn't be coddled but rather should be corralled, then indoctrinated with the ideals of public service, discipline, altruism, patriotism, and respect for authority. If Hall was somewhat provincial about how to treat adolescent storm and stress, he was nonetheless a pioneer in suggesting a biological connection between adolescence and puberty and even used language that presaged neuroscientists' later understanding of the malleability of the brain, or "plasticity." "Character and personality are taking form, but everything is plastic," he wrote, referring to pliability, not the man-made product. "Self-feeling and ambition are increased, and every trait and faculty is liable to exaggeration and excess."

Self-feeling, ambition, exaggeration, and excess—they all helped define "teenager" for the American public in the middle of the twentieth century. The teenager as a kind of cultural phenomenon took off in the post–World War II era—from teenyboppers and bobby-soxers to James Dean in *Rebel Without a Cause* and Holden Caulfield in *The Catcher in the Rye*. But while the age of adolescence became more defined and accepted, the demarcation between childhood and adulthood remained—and remains—slippery. As a society, we still carry the vestiges of our centuries-old confusion about when a person should be considered an adult. In most of the United States a person must be between fifteen and seventeen to drive; eighteen to vote, buy cigarettes, and join the military; twenty-one to drink alcohol; and twenty-five to rent a car. The minimum age to be a member of the House of Representatives is twenty-five; to be president of the United States, thirty-five; and the minimum age to be a governor ranges among states from no age restriction at all (six states) to a minimum age of thirty-one (Oklahoma). There is generally no minimum age requirement to testify in most courts, enter into a contract or sue, request emancipation from one's parents, or seek alcohol or drug treatment. But you must be eighteen to determine your own medical care or write a legally binding will, and in at least thirty-five states those eighteen or younger must have some type of parental involvement before undergoing an abortion. What a lot of mixed messages we give these teenagers, who are not at a stage to weed through the logic (if there is any) behind how society holds them accountable. Very confusing.

So what *does* being a teenager mean? Man-child, woman-child, quasi-adult? The question is about much more than semantics, philosophy, or even psychology because the repercussions are both serious and practical for parents, educators, and doctors, as well as the criminal justice system, not to mention teens themselves.

Hall, for one, believed adolescence began with the initiation of puberty, and this is why he is considered the founder of the scientific study of adolescence. Although he had no empirical evidence for the connection, Hall knew that understanding the mental, emotional, and physical changes that happen in a child's transition into adulthood could come only from an understanding of the biological mechanics of puberty.

One of the chief areas of focus in the study of puberty has long been "hormones," but hormones have gotten a bad rap with parents and educators, who tend to blame them for everything that goes wrong with teenagers. I always thought the expression "raging hormones" made it seem as though these kids had taken a wicked potion or cocktail that made them act with wild disregard for anyone and anything. But we are truly blaming the messenger when we cite hormones as the culprit. Think about it: When your three-year-old has a temper tantrum, do you blame it on raging hormones? Of course not. We know, simply, that three-year-olds haven't yet figured out how to control themselves.

In some ways, that's true of teenagers as well. And when it comes to hormones, the most important thing to remember is that the teenage brain is "seeing" these hormones for the first time. Because of that, the brain hasn't yet figured out how to modulate the body's response to this new influx of chemicals. It's a bit like taking that first (and hopefully last!) drag on a cigarette. When you inhale, your face flushes; you feel light-headed and maybe even a bit sick to your stomach.

Scientists now know that the main sex hormones—testosterone, estrogen, and progesterone—trigger physical changes in adolescents such as a deepening of the voice and the growth of facial hair in boys and the development of breasts and the beginning of menstruation in girls. These sex hormones are present in both sexes throughout childhood. With the onset of puberty, however, the concentrations of these chemicals change dramatically. In girls, estrogen and progesterone will fluctuate with the menstrual cycle. Because both hormones are linked to chemicals in the brain that control mood, a happy, laughing fourteen-year-old can have an emotional meltdown in the time it takes her to close her bedroom door. With boys, testosterone finds particularly friendly receptors in the amygdala, the structure in the brain that controls the fight-or-flight response—that is, aggression or fear. Before leaving adolescence behind, a boy can have thirty times as much testosterone in his body as he had before puberty began.

Sex hormones are particularly active in the limbic system, which is the emotional center of the brain. That explains in part why adolescents not only are emotionally volatile but may even seek out emotionally charged experiences—everything from a book that makes her sob to a roller coaster that makes him scream. This double whammy—a jacked-up, stimulus-seeking brain not yet fully capable of making mature decisions—hits teens pretty hard, and the consequences to them, and their families, can sometimes be catastrophic.

While scientists have long known *how* hormones work, only in the past five years have they been able to figure out *why* they work the way they do. Because sex hormones are present at birth, they essentially hibernate for more than a decade. What, then, triggers them to begin puberty? A few years ago, researchers discovered that puberty is initiated by what appears to be a game of hormonal dominoes, which begins with a gene producing a single protein, named kisspeptin, in the hypothalamus, the part of the brain that regulates metabolism. When that protein connects with, or "kisses," receptors on another gene, it eventually triggers the pituitary gland to release its storage of hormones. Those surges of testosterone, estrogen, and progesterone in turn activate the testes and ovaries.

After sex hormones were discovered, for the rest of the twentieth century they became the dominant theory of, and favorite explanation for, adolescent behavior. The problem with this theory is that teenagers don't have higher hormone levels than young adults—they just react differently to hormones. For instance, adolescence is a time of increased response to stress, which may in part be why anxiety disorders, including panic disorder, typically arise during puberty. Teens simply don't have the same tolerance for stress that we see in adults. Teens are much more likely to exhibit stress-induced illnesses and physical problems, such as colds, headaches, and upset stomachs. There is also an epidemic of symptoms ranging from nail biting to eating disorders that are commonplace in today's teens. We have a tsunami of input coming at teens from home, school, peers, and, last but not least, the media and Internet that is unprecedented in the history of mankind. Why are adults less susceptible to the effect of all this stimulation? In 2007, researchers at the State University of New York (SUNY) Downstate Medical Center reported that the hormone tetrahydropregnanolone (THP), usually released in response to stress to modulate anxiety, has a reverse effect in adolescents, raising anxiety instead of tamping it down. In an adult, this stress hormone acts like a tranquilizer in the brain and produces a calming effect about a half hour after the anxiety-producing event. In adolescent mice, THP is ineffective in inhibiting anxiety. So anxiety begets anxiety even more so in teens. There is real biology behind that.

In order to truly understand why teenagers are moody, impulsive, and bored; why they act out, talk back, and don't pay attention; why drugs and alcohol are so dangerous for them; and why they make poor decisions about drinking, driving, sex—you name it—we have to look at their brain circuits for answers. The elevated secretion of sex hormones is the biological marker of puberty, the physiological transformation of a child into a sexually mature human being, though not yet a true "adult."

While hormones can explain some of what is going on, there is much more at play in the teenage brain, where new connections between brain areas are being built and many chemicals, especially neurotransmitters, the brain's "messengers," are in flux. This is why adolescence is a time of true wonder. Because of the flexibility and growth of the brain, adolescents have a window of opportunity with an increased capacity for remarkable accomplishments. But flexibility, growth, and exuberance are a double-edged sword because an "open" and excitable brain also can be adversely affected by stress, drugs, chemical substances, and any number of changes in the environment. And because of an adolescent's often overactive brain, those influences can result in problems dramatically more serious than they are for adults.

2
Building a Brain

The human body is amazing, the way it neatly tucks all these complex organs into this finite space and connects them into one smoothly functioning system. Even the average human brain is said by many scientists to be the most complex object in the universe. A baby brain is not just a small adult brain, and brain growth, unlike the growth of most other organs in the body, is not simply a process of getting larger. The brain changes as it grows, going through special stages that take advantage of the childhood years and the protection of the family, then, toward the end of the teen years, the surge toward independence. Childhood and teen brains are "impressionable," and for good reason, too. Just as baby chicks can imprint on the mother hen, human children and teens can "imprint" on experiences they have, and these can influence what they choose to do as adults.

Such was the case with me. I "imprinted" on neuroscience and medicine pretty early on. My experiences cultivated in me a curiosity that I found irresistible, sustaining me from my high school years through medical school and graduate research, and to this very day. I grew up the oldest of three children in a comfortable family home in Connecticut, just forty minutes from Manhattan. I happened to live in Greenwich, which even back then was the home of actors, authors, musicians, politicians, bankers, and the independently wealthy. The actress Glenn Close was born there, President George H. W. Bush grew up there, and the great bandleader Tommy Dorsey died there.

My parents were from England; they had immigrated after World War II, and my dad came over after medical school in London to do his urological surgery residency at Columbia. To them, Greenwich seemed a great place to settle within commuting distance of New York City. It was a matter of convenience, and they were pretty oblivious to the celebrity status of the town. Perhaps because of my father, I was open-minded about learning math and science. For me a major "imprinting" moment that propelled me in the direction of medicine was a ninth-grade biology class at Greenwich Academy. The best part to me, memorable in fact, was when we each got a fetal pig to dissect. While many of my classmates slumped in their seats at the proposition of slicing up these small mammals, some rushing to the girls' washrooms with waves of nausea, a few of us jumped into the task at hand. It was one of those defining moments. The scientists had separated from those destined to be the writers, lawyers, and businesspeople of the future.

Injected with latex, the pigs' veins and arteries visibly popped out with their colorful hues of blue and red. I'm a very visual person; I also like thinking in three dimensions. That visual-spatial ability comes in handy with neurology and neuroscience. The brain is a three-dimensional structure with connections between brain areas going in every direction. It helps to be able to mentally map these connections when one is trying to determine where a stroke or brain injury is located in a patient presenting with a combination of neurological problems—definitely a plus for a neurologist. Actually, that's how the minds of most neurologists and neuroscientists work. We're a breed that tends to love to look for patterns in things. I've never met a jigsaw puzzle, in fact, that I didn't like. My attraction to neuroscience in high school and college began at a time before CT scans and MRIs, when a doctor had to imagine where the problem was inside the brain of a patient by picturing the organ three-dimensionally. I'm good at that. I like being a neurological detective, and as far as I'm concerned, neuroscience and neurology turned out to be the perfect profession for me to make use of those visuospatial skills.

If the human brain is very much a puzzle, then the teenage brain is a puzzle awaiting completion. Being able to see where those brain pieces fit is part of my job as a neurologist, and I decided to apply this to a better understanding of the teen brain. That's also why I'm writing this book: to help you understand not only what the teen brain is but also what it is not, and what it is still in the process of becoming. Among all the organs of the human body, the brain is the most incomplete structure at birth, just about 40 percent the size it will be in adulthood. Size is not the only thing that changes; all the internal wiring changes during development. Brain growth, it turns out, takes a lot of time.

And yet the brain of an adolescent is nothing short of a paradox. It has an overabundance of gray matter (the neurons that form the basic building blocks of the brain) and an undersupply of white matter (the connective wiring that helps information flow efficiently from one part of the brain to the other)—which is why the teenage brain is almost like a brand-new Ferrari: it's primed and pumped, but it hasn't been road tested yet. In other words, it's all revved up but doesn't quite know where to go. This paradox has led to a kind of cultural mixed message. We assume when someone looks like an adult that he or she must be one mentally as well. Adolescent boys shave and teenage girls can get pregnant, and yet neurologically neither one has a brain ready for prime time: the adult world.

The brain was essentially built by nature from the ground up: from the cellar to the attic, from back to front. Remarkably, the brain also wires itself starting in the back with the structures that mediate our interaction with the environment and regulate our sensory processes—vision, hearing, balance, touch, and sense of space. These mediating brain structures include the cerebellum, which aids balance and coordination; the thalamus, which is the relay station for sensory signals; and the hypothalamus, a central command center for the maintenance of body functions, including hunger, thirst, sex, and aggression.

I have to admit that the brain is not very exciting to look at. Sitting atop the spinal cord, it is light gray in color (hence the term "gray matter") and has a consistency somewhere between overcooked pasta and Jell-O. At three pounds, this wet, wrinkled tissue is about the size of two fists held next to each other and weighs no more than a large acorn squash. The "gray matter" houses most of the principal brain cells, called neurons: these are the cells responsible for thought, perception, motion, and control of bodily functions. These cells also need to connect to one another, as well as to the spinal cord, for the brain to control our bodies, behavior, thoughts, and emotions. Neurons send most of their connections to other neurons through the "white matter" in the brain. The commonly used brain imaging tool, magnetic resonance imaging, or MRI, shows the distinction between gray and white matter beautifully. On the outside surface, the brain has a rippled structure. The valleys or creases are referred to as sulci and the hills are referred to as gyri. Figure 1 shows an image from a brain's MRI scan, like those done on patients. There are two sides to the brain, each called a hemisphere. (When an MRI image shows a cut across the middle in one direction or the other [slice angles A and B], it is easier to see the two sides.) The most superficial layer of the brain is called the cortex and it is made up of the gray matter closest to the surface, with the white matter located beneath it. The gray matter is where most of the brain cells (neurons) are located. The neurons connect directly to those close by, but in order to connect to neurons in other parts of the brain, in the other hemisphere, or in the spinal cord to activate muscles and nerves in our face or body, the neurons send processes down through the white matter. The white matter is called "white" because in real life and also in the MRI scans its color is light, owing to the fact that the neuron processes running through here are coated with a fatty insulator-like substance called myelin, which truly is white in color.

As I said before, sheer size—or even weight, for that matter—doesn't mean everything. A whale brain weighs about twenty-two pounds; an elephant brain about eleven. If intellect were determined by the ratio of brain weight to body weight, we'd be losers. Dwarf monkeys have one gram of brain matter for every twenty-seven grams of body matter, and yet the ratio for humans is one gram of brain weight to forty-four grams of body weight. So we actually have less brain per gram of body weight than some of our primate cousins. It is the complexity of the way neurons are hooked up to one another that matters. Another example of how little the weight of the brain has to do with its functioning, at least in terms of intelligence, is that the

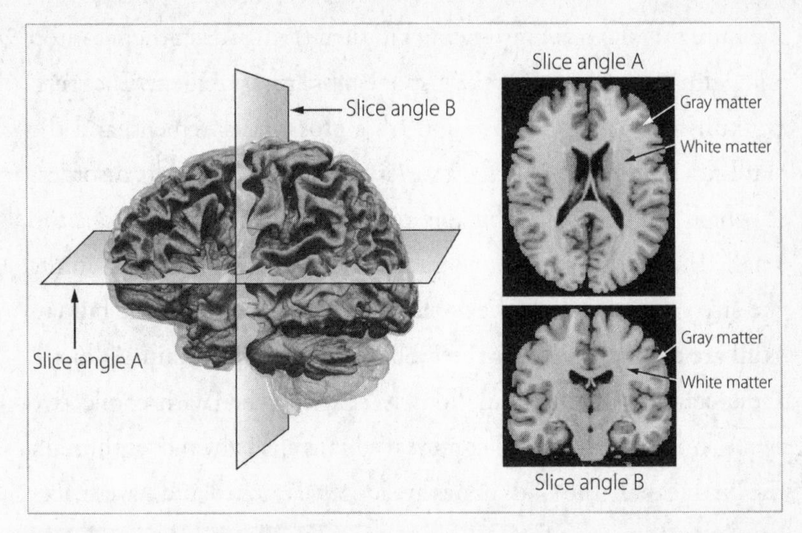

FIGURE 1. The Basics of Brain Structure: A magnetic resonance imaging (MRI) scan of a brain. The horizontal and vertical cross sections (slice angles A and B) show the cortex (gray matter) on the surface and the white matter underneath.

human female brain is physically smaller in size than the male brain but IQ ranges are the same for the two sexes. At only 2.71 pounds, the brain of Albert Einstein, indisputably one of the greatest thinkers of the twentieth century, was slightly underweight. But recent studies also show that Einstein had more connections per gram of brain matter than the average person.

The size of the human brain *does* have a lot to do with the size of the human skull. Basically, the brain has to fit inside the skull. As a neurologist, you have to measure the size of children's heads as they grow up. I have to admit there were occasions when I did this with my own sons—just like noting changes in their height—to make sure they were on track and in the normal range for skull size. When they were older, they thought I was nuts, of course, but when they were babies and toddlers, I just couldn't resist coming at them with a tape measure I'd take from my sewing kit, then trying to get them to stop squiggling free so I could take just one more measurement. The truth is, skull size doesn't tell us a lot. It's a gross measurement, and the skull can be large or small for a variety of reasons. There are disorders in which the head is too big and disorders in which the head is too small. The most important characteristic of the skull is that it limits the size of the brain. Eight of the twenty-two bones in the human skull are cranial, and their chief job is to protect the brain. At birth, these cranial bones are only loosely held together with connective tissue so that the head can compress a bit as the baby moves through the birth canal. The skull bones are loosely attached and have spaces between them: one of these is the "soft spot" all babies have at birth, which closes during the first year of life as the bones fuse together. Most growth in head size occurs from birth to seven years, with the largest increase in cranial size occurring during the first year of life because of massive early brain development.

So with a fixed skull size, human evolution did its best to jam as much brain matter inside as possible. *Homo erectus*, from whom the modern human species evolved, appeared about two million years ago. Its brain size was only about 800 to 900 cubic centimeters, as opposed to the approximately 1,500 cubic centimeters of today's *Homo sapiens*. With modern human brains nearly double the size of these ancestors', the skull had to grow as well and, in turn, the female pelvis had to widen to accommodate the larger head. Evolution accomplished all of this within just two million years. Perhaps that's why the brain's design, while extraordinarily ingenious, also gives a bit of the impression that it was updated on the fly. How else to explain the cramped conditions? Like too many clothes crammed in too small a closet, the evolution-sculpted brain looks like a ribbon repeatedly folded and pressed together. These folds, with their ridges (gyri) and valleys (sulci), as seen in Figure 1, give the human brain an irregular surface appearance, the result of all that tight packing inside the skull. Not surprisingly, humans have the most complex brain folding structure of all species. As you move down the phylogenic scale to simpler mammals, the folds begin to disappear. Cats and dogs have some, but not nearly as many as humans do, and rats and mice have virtually none. The smoother the surface, the simpler the brain.

While the brain looks fairly symmetrical from the outside, inside there are important side-to-side differences. No one is really sure why, but the right side of your brain controls the left side of your body and vice versa; this means that the right cortex governs the movements of your left eye, left arm, and left leg and the left cortex governs the movements of your right eye, right arm, and right leg. For vision, the input from the left side of the visual field goes through the right thalamus to the right occipital cortex, and information from the right visual field goes to the left. In general, visual and spatial perception is thought to be more on the right side of the brain.

The image of the body, in fact, can actually be "mapped" onto the surface of the brain, and this map has been termed the "homunculus" (Latin for "little man"). In the motor and sensory cortex, the different areas of the body get more or less real estate depending upon their functional importance. The face, lips, tongue, and fingertips get the largest amount of space, as the sensation and control necessary for these areas have to be more accurate than for other areas such as the middle of the back.

An early-twentieth-century Canadian neuroscientist, Wilder Penfield, was the first to describe the cortical map, or homunculus, which he did after doing surgery to remove parts of the brain that

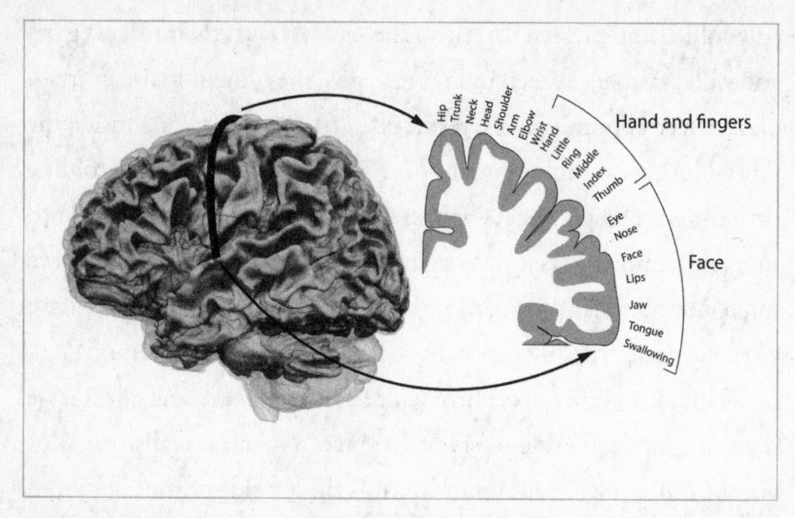

FIGURE 2. The "Homunculus": A "map" of the brain illustrating the regions that control the different body parts.

caused epileptic seizures. He would stimulate areas of the surface to determine which parts would be safe to remove. Stimulating one area would cause a limb or facial part, for instance, to twitch, and having done this on many patients he was able to create a standard map.

The amount of brain area devoted to a given body part varies depending on how complicated its function is. For instance, the area given to hands and fingers, lips and mouth, is about ten times larger than that for the whole surface of the back. (But then, what do you do with your back anyway—except bend it?) This way all the brain regions for the same part of the body end up in close proximity to one another.

My undergraduate thesis at Smith College in Northampton, Massachusetts, examined several of those areas of the brain given over to individual body parts and whether overstimulation of one of the body's limbs might result in more brain area devoted to that part.

This was actually an early experiment in brain plasticity, to see if the brain changed in response to outward stimulation. Many impressive studies that have been done since the late 1970s back up the whole concept of imprinting. Some of the most famous work, which inspired me to do my little undergraduate thesis, was done by a pair of Harvard scientists named David Hubel and Torsten Wiesel. The term that started to be used was "plasticity," meaning that the brain could be changed by experience—it was moldable, like plastic. Hubel and Wiesel showed that if baby kittens were reared with a patch on one eye during the equivalent of their childhood years (they looked sort of like pirate kittens!), for the rest of their lives they were unable to see out of the eye that had been patched. The scientists also saw that the brain area devoted to the patched eye had been partially taken over by the open eye's connections. They did another set of experiments where the kittens were raised in visual environments with vertical lines and found that their brains would respond only to vertical lines when they were adults. The point is that the types of cues and stimuli that are present during brain development really change the way the brain works later in life. So my experiment in college showed basically the same thing, not for vision, but rather for touch.

I actually had some fun showing off this cool imprinting effect in our everyday lives. Our beloved cat had died at the ripe old age of nineteen, and we all missed her so. Of course, it didn't take long before Andrew, Will, and I were at the local animal shelter, looking at kittens to bring home. We fell in love with the runt of a litter and brought home the most petite and needy little tabby kitten you could ever imagine. The boys came up with a name: Jill. Jill was always on our laps; she was a very people-friendly cat. I remembered the experiments on brain plasticity and said to Andrew and Will, when we hold her, let's massage her paws and see if she becomes a more coordinated cat. So anytime we had her in our laps, we would massage her paws with our hands, spreading them out, touching the little "fingers" that cats' paws contain. Sure enough, Jill started to use her paws much more than any other cat we had ever had (and I have had back-to-back cats since the age of eight). She used her paws for things most cats don't. She was very "paw-centric," going around the house batting small objects off tables and taking obvious pleasure in watching them hit the ground. This was a source of consternation as not all the things she knocked off were unbreakable. She also often used her left paw to eat, gingerly reaching into the cat food can with her paw and scooping up food to bring to her mouth. Watching her, we started to notice that she almost always used her left paw to do these things. We had a left-pawed cat! Then suddenly we realized that when we picked her up to massage her paws, she was facing us and because we are all right-handed, we were always stimulating her left paw much more than her right! Home neuronal plasticity demonstration project accomplished. I know if we could have looked into her brain, we'd have seen that she had more brain space given over to her paws, and especially her left paw, than the average cat. This same phenomenon of reallocating brain space based on experience during life happens in people, too. We call this part of life the critical period, when "nurture," that is, the environment, can modify "nature." But more on that later.

So what I have just told you is that brain areas for vision and body parts are compartmentalized in different places, but that they can shrink or grow relative to one another during development based on how much the senses are used. Structurally, the human brain is divided into four lobes: frontal (top front), parietal (top back), temporal (sides), and occipital (back). The brain sits on the brainstem, which connects to the spinal cord. In the rear of the brain, the cerebellum regulates motor patterning and coordination, and the occipital lobes house the visual cortex. The parietal lobes house association areas as well as the motor and sensory cortices (which include the homunculus in Figure 2). The temporal lobes include areas involved in the regulation of emotions and sexuality. Language is also located here, more specifically in the dominant hemisphere (the left temporal lobe for right-handed people and 85 percent of left-handed people, and the right temporal lobe for that small group of truly strong lefties). The frontal lobes sit most anteriorly and this area is concerned with executive function, judgment, insight, and impulse control. Importantly,

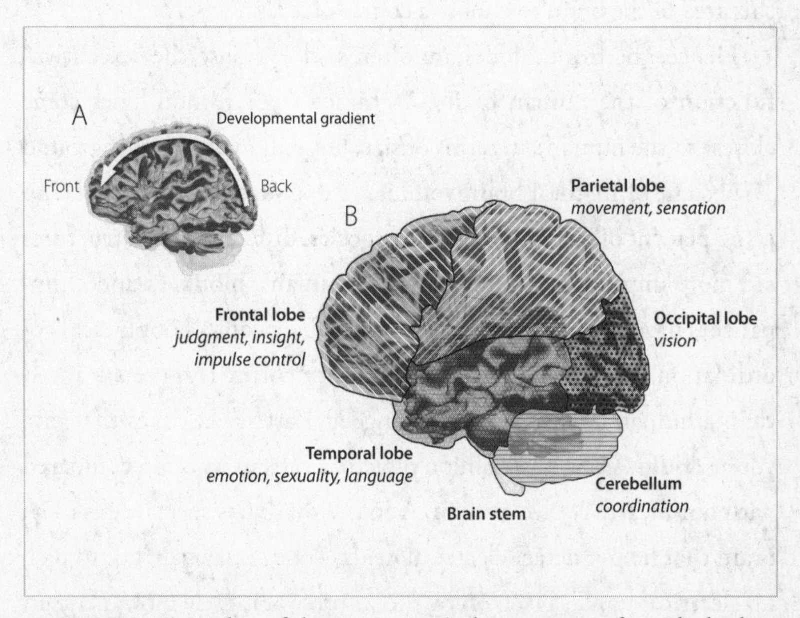

FIGURE 3. The Lobes of the Brain: A. The brain matures from the back to the front. B. The cortex of the brain can be divided into several main areas based on function.

as the brain matures from back to front in the teen years the frontal lobes are the least mature and the least connected compared with the other lobes.

The brain is divided into specialized regions for each of the senses. The area for hearing, or the auditory cortex, is in the temporal lobes; the visual cortex is in the occipital lobes; and the parietal lobes house movement and feeling in the motor and sensory cortices, respectively. Other parts of the brain have nothing to do with the senses, and the best example of this is the frontal lobes, which make up more than 40 percent of the human brain's total volume—more than in any other animal species. The frontal lobes are the seat of our ability to generate insight, judgment, abstraction, and planning. They are the source of self-awareness and our ability to assess dangers and risk, so we use this area of the brain to choose a course of action wisely.

Hence, the frontal lobes are often said to house the "executive" function of the human brain. A chimpanzee's frontal lobes come closest to the human's in terms of size, but still make up only around 17 percent of its total brain volume. A dog's frontal lobes make up just 7 percent of its brain. For other species, different brain structures are more important. Compared with humans, monkeys and chimpanzees have a much larger cerebellum, where control of physical coordination is honed. A dolphin's auditory cortex is more advanced than a human's, with a hearing range at least seven times that of a young adult. A dog has a billion olfactory cells in its brain compared with our measly twelve million. And the shark has special cells in its brain that help it detect electrical fields—not to navigate but to pick up electrical signals given off by the scantest of muscle movements in other fish as they try to hide from this deadly predator.

We humans don't have a lot else going for us other than our wile and wit. Our competitive edge is our ingenuity, brains over brawn. This edge happens to take the longest time to develop, as the connectivity to and from the frontal lobes is the most complex and is the last to fully mature. This "executive function" thus develops slowly: we certainly are not born with it!

So in what order are these brain regions all connected to one another during childhood and adolescence? This could never have been learned before the advent of modern brain imaging. New forms of brain scans, called magnetic resonance imaging (MRI), not only can give us accurate pictures of the brain inside the skull but also can show us connections between different regions. Even better, a new kind of MRI, called the functional MRI, abbreviated fMRI, can actually show us what brain areas turn one another on. So we can actually see if areas that "fire" together are "wired" together. In the last decade, the National Institutes of Health conducted a major study to examine how brain regions activate one another over the first twenty-one years of life.

What they found was remarkable: the connectivity of the brain slowly moves from the back of the brain to the front. The very last places to "connect" are the frontal lobes (Figure 4). In fact, the teen brain is only about 80 percent of the way to maturity. That 20 percent gap, where the wiring is thinnest, is crucial and goes a long way toward explaining why teenagers behave in such puzzling ways—their mood swings, irritability, impulsiveness, and explosiveness; their inability to focus, to follow through, and to connect with adults; and their temptations to use drugs and alcohol and to engage in other risky behavior. When we think of ourselves as civilized, intelligent adults, we really have the frontal and prefrontal parts of the cortex to thank.

Because teens are not quite firing on all cylinders when it comes to

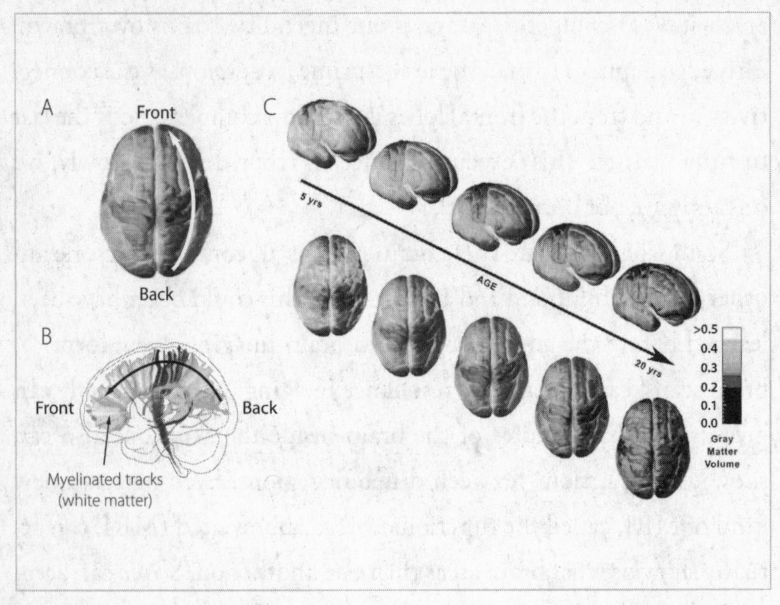

FIGURE 4. Maturing Brain: The Brain "Connects" from Back to Front: A. A functional MRI (fMRI) scan can map connectivity in the brain. Darker areas indicate greater connectivity. B. Myelination of white matter tracks cortex maturation from back to front; this is why the frontal lobes are the last to be connected. C. Serial connectivity scans reveal that frontal lobe connectivity is delayed until age twenty or older.

the frontal lobes, we shouldn't be surprised by the daily stories we hear and read about tragic mistakes and accidents involving adolescents. The process is not really done by the end of the teen years—and as a result the college years are still a vulnerable period. Recently a friend of mine told me about his son's college classmate, Dan, an all-around great kid who'd rarely caused his parents to worry. He was popular, had been a star ice hockey player in high school, and was a finance major in college. Over the summer my friend's son got a phone call from Dan's mother. Dan had drowned the night before, she told him. He'd been out with friends, drinking, and sometime between three and four in the morning, on their way home, the group—there were eight of them—decided they wanted to cool off, so they stopped at the local tennis club. The club was closed, of course, but the locked gate didn't stop them. All eight scaled the fence and jumped into the pool. It was only after they'd gotten home that someone said, "Where's Dan?" Racing back to the club, they found their friend facedown in the water. The medical examiner listed the cause of death as accidental drowning due to "acute alcohol intoxication." One of the news reports I read made me shake my head: "Police are asking kids and adults to think twice about potential dangers before taking any risks that could turn deadly."

"Think twice."

How many times have we all said this to our teenage sons and daughters? Too many times. Still, as soon as I heard about Dan, I called my boys to tell them the story. You have to remember this, I told them. This is what happens. Drinking and swimming don't go together. Neither does the decision to suddenly scale a fence in the middle of the night, or jump into a pool with seven friends who are also intoxicated.

How parents deal with these tragic stories and talk about them with their own kids is critical. It shouldn't be, "Oh, wow, I'm so glad that wasn't my child." Or, "My teenager would never have done that." Because you don't know. Instead, you have to be proactive. You have to stuff their minds with real stories, real consequences, and then you have to do it again—over dinner, after soccer practice, before music lessons, and, yes, even when they complain they've heard it all before. You have to remind them: These things can happen anytime, and there are many different situations that can get them into trouble and that can end badly.

One of the reasons that repetition is so important lies in your teenager's brain development. One of the frontal lobes' executive functions includes something called prospective memory, which is the ability to hold in your mind the intention to perform a certain action at a future time—for instance, remembering to return a phone call when you get home from work. Researchers have found not only that prospective memory is very much associated with the frontal lobes but also that it continues to develop and become more efficient specifically between the ages of six and ten, and then again in the twenties. Between the ages of ten and fourteen, however, studies reveal no significant improvement. It's as if that part of the brain—the ability to remember to do something—is simply not keeping up with the rest of a teenager's growth and development.

The parietal lobes, located just behind the frontal lobes, contain association areas and are crucial to being able to switch between tasks, something that also matures late in the adolescent brain. Switching between tasks is nearly a constant need in today's world of information overload, especially when you consider the fact that multitasking—doing two cognitively complex things at the same time—is actually a myth. Chewing gum and doing virtually anything else is not multitasking because chewing gum involves no real cognitive focus. Both talking on a cell phone and driving, however, do involve cognitive focus. Because there are limits to how many things the human brain can focus on at any one time, when someone is engaged in multiple cognitively significant activities, like talking and driving, the brain must constantly switch back and forth between the two tasks. And when it does, neither of those tasks is being accomplished particularly well.

The parietal lobes help the frontal lobes to focus, but there are limits. The human brain is so good at this juggling that it seems as though we are doing two tasks at the same time, but really we're not. Scientists at the Swedish medical university Karolinska Institutet measured those limits in 2009 when they used fMRI images of people multitasking to model what happens in the brain when we try to do more than one thing at a time. They found that a person's working memory is capable of retaining only between two and seven different images at any one time; this means that focusing on more than one complex task is virtually impossible. Focusing chiefly happens in the parietal lobes, which dampen extraneous activity to allow the brain to concentrate on one thing and then another.

The problem of having immature parietal lobes was illustrated in a segment on *Good Morning America* in May 2008 by the ABC TV correspondent David Kerley and his teenage daughter Devan. Using a course set up by Allstate Insurance, and with her father in the passenger seat, Devan, who had been driving for a year, was instructed about speed, braking, and turning and allowed to take a practice run through the course. Then she was given a series of three "distractions" to handle while navigating the course's twists and turns. First, she was handed a BlackBerry and told to read the text on the screen while driving. She hit several cones. Next, three of her friends were put in the backseat and a lively conversation ensued. Devan hit more cones. Finally, Devan was handed a package of cookies and a bottle of water, and just passing the cookies around and holding the bottle of water caused her to run over several more cones. Multitasking is not only a myth but a dangerous one, especially when it comes to the teenage brain.

"Multitasking" has become a household word. The research in Sweden suggests that there are limits. Teenagers and young adults pride themselves on their ability to multitask. Have today's teens and young adults imprinted on a multitasking world? Maybe. In studying how young adults these days handle distractions, researchers at the University of Minnesota have shown that the ability to successfully switch attention among multiple tasks is still developing through the teenage years. So it may not come as a surprise to learn that of the nearly six thousand adolescents who die every year in automobile accidents, 87 percent die because of distracted driving.

The question of whether today's teens and young adults have a special skill set for learning while distracted was more formally tested in 2006 by researchers at the University of Missouri. They took twenty-eight undergraduates, including kids in their late teens, and asked them to memorize lists of words and then recall these words at a later time. To test whether distraction affected their ability to memorize, the researchers asked the students to perform a concurrent task—placing a series of letters in order based on their color by pressing the keys on a computer keyboard. This task was given under two conditions: when the students were memorizing the lists of words and when the students were recalling those lists for the researchers. The Missouri scientists discovered that simultaneous tasks affected both encoding (memorizing) and retrieving (recalling). When the keyboard task was given while the students were trying to recall the previously memorized words (which is akin to taking a test or exam), there was a 9 to 26 percent decline in their ability to memorize the words. The decline was even more if the concurrent distracting task occurred while they were memorizing, in which case their performance decreased by a whopping 46 to 59 percent.

These results certainly have implications for the teen bedroom during a homework night! I not-so-fondly remember walking in on my sons during evening homework time to find them with the television on, headphones attached to iPods, all the while messaging someone on the lower corner of their computer screens and texting someone else on their iPhones. It wasn't a problem, they protested, when I suggested they concentrate on their homework, assuring me their course reviews for the next day's exams were totally unaffected by the thirty-two other things they were doing at the same time. I didn't buy it. So I buttressed my argument with the Missouri data. I put Figure 5 in this book in case you want to use it to make the same point to your teen.

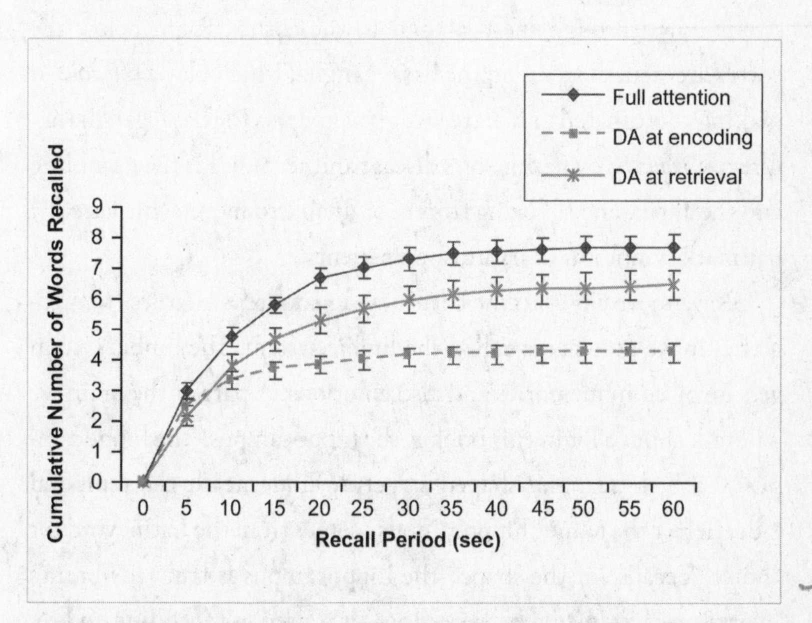

FIGURE 5. Multitasking Is Still Not Perfect in the Teen Brain: College students were tested under three conditions: No Distraction (full attention), Distracted Attention (DA) when memorizing (DA at encoding), and Distracted Attention when recalling (DA at retrieval). Students performed poorly when multitasking during recall, and even worse when they multitasked while memorizing.

Attention is only one way we can assess how the brain is working. There's a lot more under the hood of the brain than just the four lobes, so returning to Figure 3 let's start at the back, where we find the brainstem at the very bottom of the brain, attached to the spinal cord. The brainstem controls many of our most critical biological functions, like breathing, heart rate, blood pressure, and bladder and bowel movements. The brainstem is on "automatic"—you are not even aware of what it does, and you normally don't voluntarily control what it does. The brainstem and spinal cord are connected to the higher parts of the brain through way station areas, like the thalamus, which sits right under the cortex. Information from all the senses flows through the thalamus to the cortex. Right below the cortex are structures called the basal ganglia, which play a big role in making coordinated and patterned movements. The basal ganglia are directly affected by Parkinson's disease and account for the trembling and the appearance of being frozen, or unable to move, which are the hallmark symptoms of Parkinson's patients.

As we move closer to the cortex, we encounter structures that together make up what is called the limbic system. The limbic system gets involved in memories and also emotions. A part of the brain we will talk about a lot in this book is the hippocampus. The hippocampus is a little seahorse-shaped structure underneath the temporal lobe. In fact the name "hippocampus" comes from the Latin word for "horse" because of the shape. The hippocampus is truly the brain's "workhorse" for memory processing—it is used for encoding and retrieving memories.

So what do we know about our memory workhorse? It has the highest density of excitatory synapses in the brain. It is a virtual beehive of activity, and turns on with every experience. As we will explain later, the hippocampus in the adolescent brain is relatively "supercharged" compared with an adult's.

The connection of the hippocampus to memory was recognized some six decades ago through the unforeseen consequences of one patient's radical brain surgery. This surgery was performed in 1953 on a twenty-seven-year-old Connecticut man who, until his death several years ago, was known only by his initials, H.M. He underwent an experimental operation in an attempt to cure him of frequent and severe epileptic seizures. So incapacitating was H.M.'s epilepsy that he was unable to hold down even a factory job. When the Yale neurosurgeon William Beecher Scoville removed most of H.M.'s medial temporal lobe, which was causing his seizures, the operation appeared to be a success. By cutting away brain tissue in the area of the seizures, Scoville dramatically reduced their frequency and severity. In the process, though, he also removed a large portion of H.M.'s hippocampus. (That the hippocampus is critical for memory formation was unknown at the time; the case of H.M. shed much light on the subject.) What became clear when H.M. awoke was that while his seizures were by and large gone, so, too, was his ability to turn short-term memories into long-term memories. Essentially, H.M. could remember his past—everything before the time of the operation—but for the rest of his life he had no short-term memory and could not remember what happened to him, what he said or did or thought or felt or whom he met, in the decades following the surgery. H.M.'s loss, as often happens in the history of science, was neuroscience's gain. For the first time researchers could point to a specific brain region (the temporal lobe) and brain structure (the hippocampus) as the seat of human memory.

Next door to the hippocampus, in another part of the limbic

system under the temporal lobe, is another key brain structure, the amygdala, which is involved in sexual and emotional behavior. It is very susceptible to hormones, such as sex hormones and adrenaline. It is sort of the seat of anger, and when stimulated in animal experiments, it has been shown to produce rage-like behavior. The limbic system can be thought of as a kind of crossroads of the brain, where emotions and experiences are integrated.

A slightly unbridled and overexuberant immature amygdala is thought to contribute to adolescent explosiveness; this explains in part the hysteria that greets parents when they say no to whatever it is their adolescent thinks is a perfectly reasonable request. Cross that immature amygdala with a teen's loosely connected frontal lobe, and you have a recipe for potential disaster. For example, the sixteen-year-old patient of a colleague of mine was so incensed when his parents said driving was a "privilege" (for which he did not yet qualify), and not a "right," that he stole the car keys and drove away from the house. He didn't get very far, though. He forgot the garage door was closed and plowed right through it. One of my colleagues also told me that, because he himself had three grown daughters, rather than sons, he had few "terrible teen tales" to tell. Then he reconsidered: "Oh, yes, there was the weekend we were away and the 'couple of friends' became a party that got out of hand, including the raid on our wine cellar, a minor fender bender with our stolen liquor in the trunk, and maybe a navel ring (which I never knew about until years later after it disappeared). But all's well that ends well."

3
Under the Microscope

If you pick out any random region of brain and look at it under a microscope, you'll find it jam-packed with cells. In fact, there is almost no space between the billions of cells in the brain. Evolution made sure of that, putting to use every cubic micron wisely. A cell is the body's smallest unitary building block, and each has its own command center, called a nucleus, a large oval body near the center of the cell. There are more than two hundred different types of cells making up every organ, tissue, muscle, etc. A unique cell type in the brain is the neuron. This is a cell we will talk about frequently in this book. Thoughts, feelings, movements, and moods are nothing more than neurons communicating by sending electrical messages to one another.

I remember my first time looking at brain cells under a microscope. In the mid- to late 1970s the only way to study changes in neurons, for instance the changes that occur during learning, was by looking through a microscope at individual cells over a given period of time. Today, we have amazing tools—brain imaging scans and specialized microscopes—that allow us to look into the brain and see cells and synapses change in real time. If you are learning something right now, as you read this, your neurons will change in about fifteen minutes, creating more synapses and receptors. Changes start within milliseconds of learning something new, and can take place over a period of minutes and hours. When I look at brain cells under a microscope, I think of the billions of neurons that are interconnected and how we're still trying to figure out the wiring. What we know now is that no two human brains are wired exactly the same, and experience shapes us all differently. It's the final frontier, our own internal frontier, and we're just now beginning to see all the patterns.

There are 100 billion neurons in the human brain and you could place about 30,000 of them on the head of a pin, but placed end to end the neurons in just one person's cortex would stretch for 100,000 miles—enough to circle the globe four times. At birth, we have more neurons than at any other time in our life. In fact, our brains are at their densest before birth, between the third and sixth months of gestation. Dramatic pruning of much of that gray matter occurs in the last trimester and first year of life. Still, by the time a baby is born, he or she has a brain brimming with neurons. Why? An infant's overabundance of neural cells is needed to respond to the barrage of stimuli that comes with entering into the world. In response to all those new sights, sounds, smells, and sensations, neurons branch out in the baby's brain, creating a thick forest of neural connections. So why aren't all babies tiny Mozarts and Einsteins? Because when we are born, only a very small percentage of that overflow of neurons is wired together. The information is going in, being absorbed by the neurons, but it doesn't know where to go next. Like someone plunked down in the middle of a strange and bustling metropolis, the infant brain is surrounded with possibilities and yet has no map, no compass, to navigate this strange new world. "All infants are born in a state of psychedelic splendor similar to an acid trip" is how Daniel Levitin, a neuroscientist at McGill University in Montreal, Canada, colorfully describes it.

A neuron responds to a stimulus with a burst of activity, called an action potential, which is actually an electrical signal that passes, in relay fashion, from the point of contact with the stimulus down the receiving limb of the neuron, called the dendrite, through the cell.

When we see the color red, smell a rose, move a muscle, or remember someone's name, action potentials are happening.

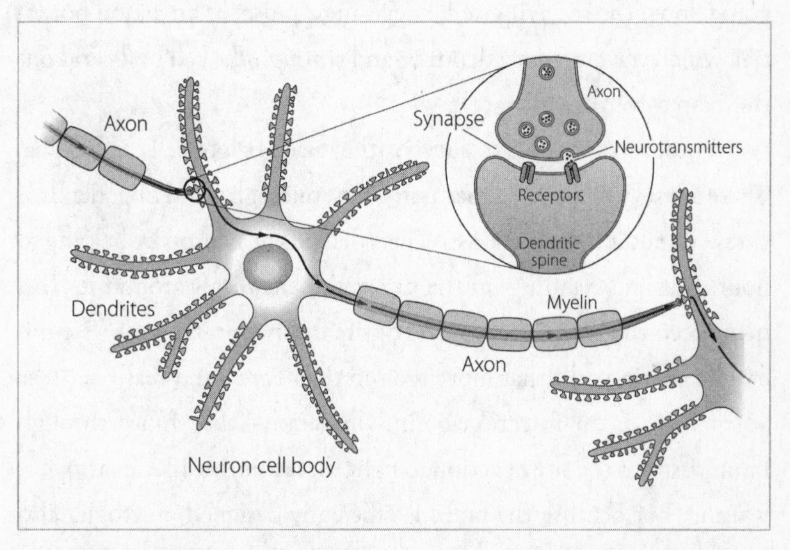

FIGURE 6. Anatomy of Neuron, Axon, Neurotransmitter, Synapse, Dendrite, and Myelin: Signals between cells flow in one direction, from an axon to a dendritic spine, through a synapse. Axons with myelin coating transmit signals faster than those without. At the synapse, a transfer of neurotransmitter molecules binds to the synaptic receptor on the spine.

If you think of each neuron's cell body as a point in a relay, there must be an incoming and an outgoing signal. Once the outgoing signal reaches the axonal bouton, or end point, it sets off a reaction causing the bouton to release packages of chemical messengers, called neurotransmitters. The point of contact between two neurons is called a synapse, and is actually a space no more than two-millionths of an inch wide. The synapse is truly where the action takes place in the brain. The signal heads down the neuron through the axon to the synapse and is then released as a chemical message. Like liquid keys, these neurotransmitters cross the synapse and lock onto the neuron on the other side, and in this way carry information from one cell to another. Once opened, the receptor causes a chain reaction of signals going down the receiving cell, triggering a pulse, or an action potential, which travels from a dendrite and through the cell body and out the axon toward another cell.

In order for neurons to survive, they need helper cells called glia. There are several types of glia: astrocytes, microglia, and oligodendrocytes. To put it simply, the astrocytes defend the neuron by helping to nourish it and cleaning up the unwanted chemicals around it. This helps keep the brain's neurons at optimal functioning level. The microglia are tiny cells that move around the neuron and really activate when there is an infection or inflammation—they move through brain tissue to the site of action to fight these injuries, like an army-in-waiting. But because the brain is efficiently designed, microglia also have an everyday purpose, a kind of housekeeping duty, so that even when they are not activated, they are still helping maintain the health and well-being of the synapses. Oligodendrocytes are the cells that make the myelin that goes around the axon of neurons. These cells are tightly packed in the white matter, wrapping whitish-colored myelin around axons to insulate them, much like rubber around an electrical cord, allowing faster speeds of signal transmission down the axon.

While you are born with the vast majority of your neurons, most of the synapses in the cortex are not fully formed. In lower areas, like the brainstem, synapses are indeed almost fully mature. In the cortex, however, synapses are produced after birth in a burst of activity, which I mentioned earlier, known as the critical period. During this stage of development, a baby's brain creates an astonishing two million synapses every second, allowing the infant to reach mental milestones like color vision, grasping, facial recognition, and parental attachment. It's as if an infant's brain is sending out billions of antennae, scanning the world for information. For each synapse to survive, it must find another neuron to send information to; this is why the number of synapses in a baby's brain peaks in childhood. The gray matter—the brain tissue responsible for processing information—continues to thicken throughout childhood as the brain's cells form extra connections, those limb-like dendrites. Known as arborization, this thickening is like a tree growing extra branches and roots. Stimulation, experiences, repeated sensations—all contribute to the creation of these new neural pathways. In adolescence, this "overgrowth" is responsible for a teen's heightened capacity to learn new things quickly—everything from operating the new TV remote to speaking Mandarin Chinese. The profusion of gray matter, though, can also cause a kind of cognitive dissonance in which the brain has trouble picking out the right signals from all the "noise." As a result, by late adolescence the brain has begun to prune away excess synapses and streamline connections.

Synapses come in two flavors: ones that excite, or turn on, the next neuron, and ones that inhibit, or turn off, the next neuron.

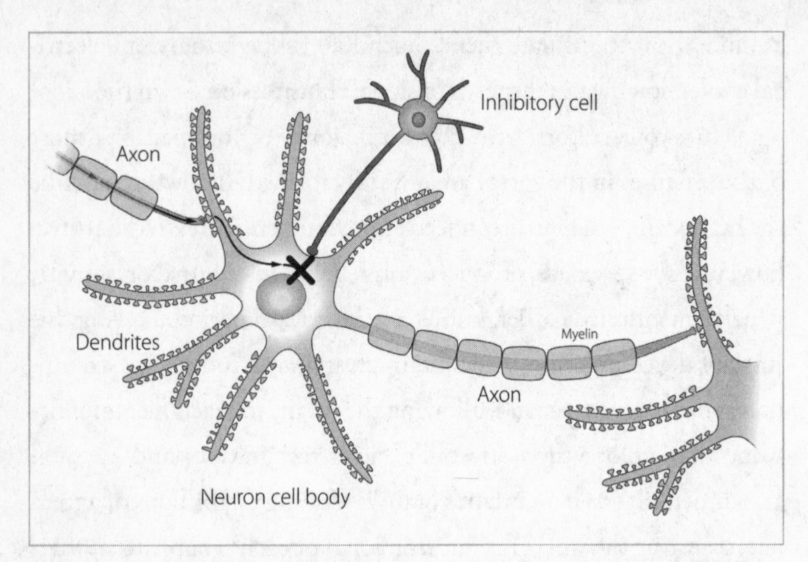

FIGURE 7A. Inhibitory Cells Can Stop Signaling: Inhibitory cells release inhibitory neurotransmitters onto spines, which will stop a signal in a neuron and turn the cell "off."

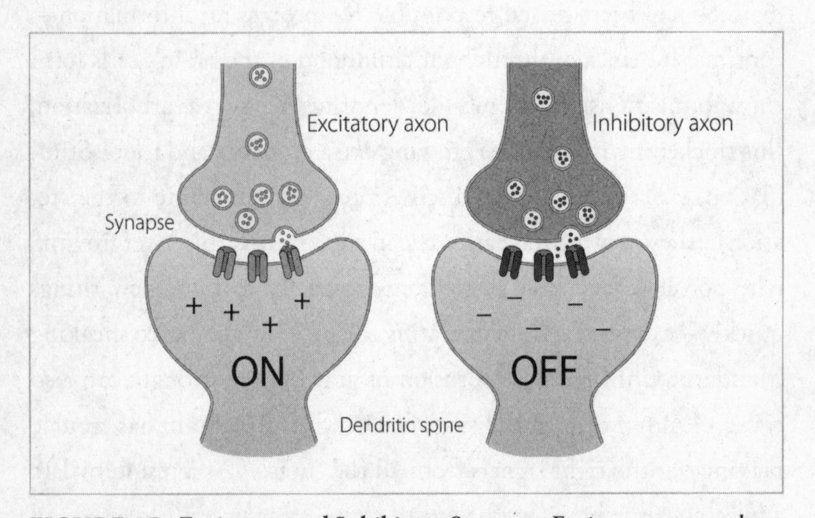

FIGURE 7B. Excitatory and Inhibitory Synapses: Excitatory axons release excitatory neurotransmitters, such as glutamate, which bind to excitatory receptors and turn the neuron "on." Inhibitory axons release inhibitory neurotransmitters, like GABA, which bind to inhibitory receptors and turn the neuron "off."

Whether or not the synapse is excitatory or inhibitory depends upon the type of neurotransmitter the axon puts out and also on the custom-made receptor, or lock, which is the part of the synapse poised to "receive" the neurotransmitter. If you imagine the neurotransmitter as a simple geometric shape, say a square or a circle, the specific receptor for that "flavor" of neurotransmitter will have the complementary shape in order to make a perfect fit. Just as "you can't put a square peg in a round hole," these neurotransmitter "keys" will fit into only the perfect receptor "locks." This helps the synapse not confuse messages. In addition to the near-perfect pairing of neurotransmitters to receptors, another way the signal is kept clean is that the astrocyte helper cells immediately clean up any leftover neurotransmitter hanging around after it gets released. This happens in milliseconds, as the timing of these signals between brain cells has to be rapid, sharp as a burst.

Once the neurotransmitter has bound and locked itself into the receptor on the receiving neuron, this pairing sets off a chain reaction. Inside the dendritic side of the synapse, there are lots of proteins that rush to work when the synapse gets excited or inhibited. The signal needs to get down the dendrite to the cell body of the neuron, where it sends a positive charge for an excitatory signal or a negative charge for an inhibitory signal. Depending on which charge is sent, the receiving neuron will get a message to either stop or start functioning. If the message is positive, the receiving neuron will send the information down its own axon and across another synaptic cleft, and so on. A neuron can have up to ten thousand synapses and can send a thousand electrical impulses every second. In one-tenth of the time it takes to blink your eyes, a single neuron can simultaneously send a signal to hundreds of thousands of other neurons.

Some of the most common excitatory neurotransmitters are epinephrine, norepinephrine, and glutamate. Inhibitory neurotransmitters, like gamma-aminobutyric acid (GABA) and serotonin, act as antianxiety nutrients, calming the body and telling it to slow down. A lack of serotonin can result in aggression and depression. Dopamine is a special neurotransmitter because it is both excitatory and inhibitory. It is also, along with epinephrine and several others, a hormone. When it acts on the adrenal glands, it is acting hormonally; when it acts in the brain, it is a neurotransmitter. As a brain chemical messenger, dopamine helps motivate, drive, and focus the brain because it is integral to the brain's reward circuitry. It's the "I gotta have it" neurochemical that not only reinforces goal-directed activity but also can, in certain circumstances, lead to addiction. The more dopamine that is released in the brain, the more the reward circuits are activated, and the more those circuits are activated, the bigger the craving. It doesn't matter if the craving is at the dinner table or the card table, in the boardroom or the bedroom. For instance, scientists know that high-calorie foods produce more dopamine in the brain. Why? Because higher calories increase our chance for survival. When we crave ice cream or gambling or sex, we may not actually be craving sweets, money, or orgasms. We're craving dopamine.

Inhibiting a neural response is just as important as activating one when it comes to "executive" brain function. Examples of things that bind to inhibitory synapses are sedatives such as barbiturates, alcohol, and antihistamines. Synapses will be critical in our discussion of the adolescent brain because both the number and the type of synapses in our brains change as we age. They also change in relation to the amount of stimulation our brains experience. One topic that will come up later is the effect of illegal and illicit drugs and alcohol on these synapses, which we will cover in the chapter on addiction.

A popular instrument used by researchers to test inhibition is the Go/No-Go task in which subjects are told to press a button (the "Go" response) when a certain letter or picture appears, and *not* to press it (the "No-Go" response) when the letter *X* appears. Several studies have shown that children and adolescents generally have the same accuracy, but the reaction times, the speed at which a subject successfully inhibits a response, dramatically decrease with age in subjects age eight to twenty. In other words, it takes longer for adolescents to figure out when *not* to do something.

Signals move from one area of the brain to another along fiber tracks, and some of these tracks travel down through the core regions of the brain in order to send signals to and from the spinal cord. Brains are intricately interconnected by these fibers, and research using special brain scans is rapidly evolving to look at these connections. Because axons are designed to have a rapid pulse of electricity run through them to the connection point at the synapse, they act like electrical wires conducting an electrical signal. And just as an electrical wire needs insulation in order for the electricity not to dissipate along its length, so do the axons. Since we don't have rubber in our brains, our axons are coated with a fatty substance called myelin. (See Figure 6.) The brain requires myelin in order to function normally, to get a signal from one region of the brain to another and also down to the spinal cord. As we said before, myelin is made by oligodendrocytes, and has a white hue due to its fatty content: hence the term "white matter." By essentially "greasing" the "wires," myelin allows signals to travel down axons faster, increasing the speed of a neural transmission as much as a hundredfold. Myelin also aids the speed of transmission by helping to cut down the synapses' recovery time between neural firings, thereby allowing a thirtyfold increase in the frequency with which neurons transmit information. The combination of increased speed and decreased recovery time has been estimated by researchers as roughly equivalent to a three-thousand-fold increase in computer bandwidth. (Myelin also is the target of attack in the disease multiple sclerosis, or MS. Patients with MS have areas of inflammation in their white matter that come and go, and this is why they can lose functions like walking, sometimes only temporarily until the inflammation passes.)

At birth, a baby's cortex contains little myelin; this explains why the electrical transmissions are so sluggish and an infant's reaction times so slow. However, the baby's brainstem is almost as fully myelinated as an adult's, so it can control automatic functions like breathing, heartbeat, and gastrointestinal function necessary to stay alive. Connections to and from many other areas of the brain occur after birth, beginning with the motor and sensory areas at the bottom and back of the brain. As these areas become wired with myelin, infants are better able to process basic information from their senses—their eyes, ears, mouth, skin, and nose. Within the first year, the neural tracts that support brain regions involved in vision and other primary senses, as well as those involved in gross motor activity, are completed. This is, in part, why it takes about a year for a baby to become coordinated enough to walk. Much of the brain becomes insulated by age two, and high-level areas involved in language and fine motor coordination follow over the next few years when children are particularly primed to learn to talk and improve their fine motor skills. The more complex areas of the brain, especially the frontal lobes, take much, much longer and are not finished until a person is well into his or her twenties.

All of this learning is dependent on excitation, the driving force in our brains. Excitatory signals between neurons build brain connections and are required for brain development. Excitation can come from outside or inside your brain, but regardless, if a particular pathway of cells and their synapses are activated repeatedly, the synapses between them strengthen. Thus, cells that "fire" together "wire" together.

In the developing brain, especially in early childhood, as groups and pathways of neurons and their synapses get activated, the process of excitation "turns on" the molecular machinery in the cell. This actually results in the building of more synapses, a process we term synaptogenesis (birth of synapses). Synapses are increased in infancy through adolescence, peaking in early childhood. Because synaptogenesis is so dependent upon brain cells being activated by one another, a child's brain has more excitatory than inhibitory neurotransmitters and synapses compared with an adult's brain, where there is more balance between the two.

Excitation is a key element of learning. The period in early life in which excitation is so prominent is also called the "critical period," when learning and memory are more robust than in later life. This allows the brain to be very sensitive to excitation and grow. Unfortunately, the abundant excitation in the developing brain carries a price: the risk for overexcitation. This explains why diseases that are a result of overexcitation, like epilepsy, are more common in childhood than adulthood. Seizures are the main symptom in epilepsy, and they are

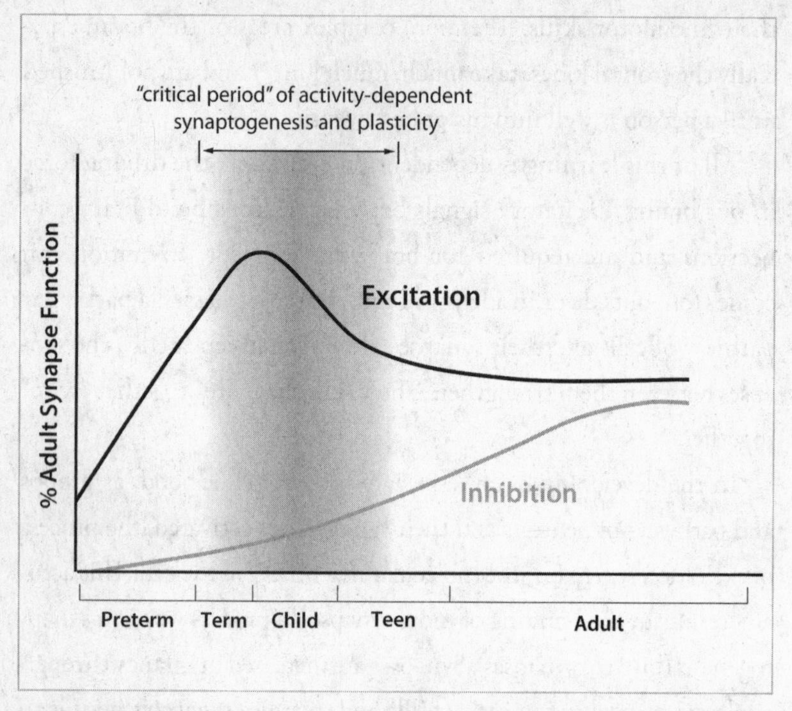

FIGURE 8. The Young Brain Has More Excitatory Synapses Than Inhibitory Synapses: The number of synapses increases from infancy through adolescence, peaking in early childhood.

caused by too many brain cells turning on together without enough inhibition to balance them.

Arborization, or the branching out of neurons, peaks in the first few years of life but continues, as we've seen, into adolescence. Gray matter density peaks in girls at age eleven and in boys at age fourteen, and waxes and wanes throughout adolescence.

White matter, or myelin, however, has only one trajectory in adolescence: up. Jay Giedd and colleagues at the National Institute of Mental Health scanned the brains of nearly one thousand healthy children, ages three to eighteen, and discovered this pattern of wiring.

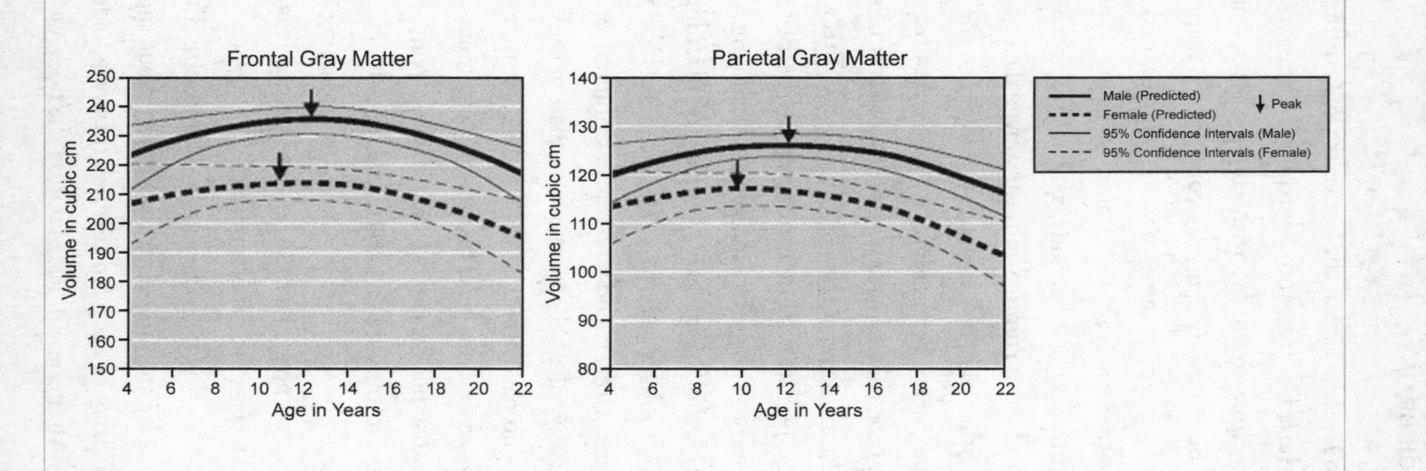

FIGURE 9. Gender Differences in Rate of Cortical Gray Matter Growth: Like the body, the male brain is on average larger than the female brain. Rates of growth in male and female brains also are different. In females, the growth rate of two areas important for cognitive maturity—the frontal lobes and the parietal lobes—peaks in the early teen years, but in males the peak does not occur until the late teens.

As we saw in Figure 4, researchers at the University of California, Los Angeles, built on those findings and compared the scans of young adults, ages twenty-three to thirty, with those of teenagers, ages twelve to sixteen. They found that myelin continues to be produced well past adolescence and even into a person's thirties, making the communication between brain areas ever more efficient.

Without those insulated connections, a signal from one area of the brain, say fear and stress coming from the amygdala, has trouble linking up with another part of the brain, for instance the frontal cortex's sense of judgment. For adolescents whose brains are still being wired, this means they sometimes find themselves in dangerous situations, not knowing what they should do next. This was confirmed scientifically in a 2010 study conducted by the British Red Cross into how teenagers react to emergencies involving a friend drinking too much alcohol. More than 10 percent of all children and young teens between the ages of eleven and sixteen have had to cope at one time or another with a friend who was sick, injured, or unconscious owing to excessive alcohol consumption. Half of those had to deal with a friend who passed out. More broadly, the survey found that nine out of ten adolescents have had to deal with some kind of crisis involving another person during their teenage years—a head injury, choking, an asthma attack, an epileptic seizure, etc. Forty-four percent of the teens surveyed admitted to panicking in that emergency situation, and nearly half (46 percent) acknowledged they didn't know how to respond to the crisis at all.

Dan Gordon, a fifteen-year-old boy from Hampshire, England, who was interviewed by the *Guardian* for a story about the study, spoke about a house party he attended at which there was widespread underage drinking. After one girl passed out on the floor, facedown, she began to vomit, and the others in the room, all teenagers, panicked. Thinking only that they needed to prevent her from choking, they stood her up and, with effort, walked her outside for fresh air and waited for her to wake up. Dan admitted to the reporter that neither he nor anyone else at the party had thought to call for an ambulance. In other words, the teenagers' amygdalae had signaled danger, but their frontal lobes didn't respond. Instead, the teens acted *in the moment*.

My son Andrew witnessed something similar during college. He was visiting his then-girlfriend at a college in Boston. The girlfriend's roommate also had an out-of-town visitor, a shy freshman girl from the South who quickly became intoxicated at a party in another student's room. When Andrew and his girlfriend returned to her dorm, they found the young girl passed out, and just as in Dan Gordon's story, they all panicked. Instead of calling 911 or campus security, or driving her to an emergency room, they found a couple of friends to help, and then drove all the way out to our house, about ten miles away.

"We didn't want to call campus security," Andrew's girlfriend explained, as I observed the young girl, whom they had helped into the house and who was now almost unresponsive. "She's a freshman. If we brought her to the health center, me and my roommate could get in trouble."

Andrew and his former girlfriend were both twenty-one at the time, but the visiting student was just eighteen.

"What about taking her to the hospital?" I said.

"We didn't know how drunk she was," the other friend said. "She was talking when we put her in the car, and now she's completely out of it."

None of them in fact knew the girl—they had met her briefly for the first time earlier that day, when she had arrived to visit the roommate. She had her wallet and an ID from her South Carolina college with her, but no other information. The roommate who had invited her to Boston was nowhere to be found. Already drowsy, she was rapidly becoming more sedated, and then she vomited on the floor. At that point, I insisted they get her to a local community hospital just a mile from our house. It took three of them to half-carry her back to the car. About fifteen minutes later, I got a call from Andrew's girlfriend, who said the hospital was going to admit the girl for observation. The poor thing spent an unhappy night in the hospital, and the college crew picked her up the next afternoon. On their way back to Boston, they stopped by my house to gather things they had left there the night before. The young freshman looked pale and very tired, but otherwise was fine. Apparently her blood alcohol level had peaked at 0.34, which was more than four times the legal driving limit, and life threatening. Had she not been taken to the hospital, where her stomach was pumped and charcoal administered to prevent her body from absorbing any more alcohol, I shudder to think of what might have happened. As I had a captive audience, I sat them all down in the kitchen, turned on my laptop, and showed them a chart about blood alcohol levels and the effects on coordination and consciousness. I pointed out that 0.4, which was only a little more than her blood alcohol reached at its height, can be lethal. Turns out she had done about seventeen Jell-O shots that evening—to the best of her memory. There was no point in asking the usual question—"What were you thinking?"—but I felt this was a good teaching moment to show them all how close she had come to a very different end the night before.

The young girl recovered and hopefully learned her lesson, but obviously the consequences of poor decision-making can be, and often are, disastrous for teens. Bennett Barber was sixteen years old on New Year's Eve 2008 when he left an unsupervised party at a friend's house in Marblehead, Massachusetts, and began to walk home. It was around 11:30 p.m., snow was falling, and the wind was gusting up to thirty miles per hour. Dressed in jeans and sneakers, Bennett was drunk and disoriented, and although his home was only a half mile away, he became lost. With the temperature plunging into single digits, Bennett eventually collapsed, face-first, into a snowbank. At three o'clock in the morning his mother notified the police, and a search party was sent out into the subfreezing night. Hours later, a firefighter discovered a beer bottle in the snow and followed a blurry set of footprints. When he found Bennett, the boy was semiconscious and suffering from hypothermia. He was also missing a sneaker and a sock. The high school hockey player was taken by ambulance to Massachusetts General Hospital, where his core temperature was only 88 degrees and his right foot appeared to be frozen solid. Isolated in a special chamber to raise his body temperature, he was eventually transferred to a burn center for treatment of his frostbite.

Bennett later told his father why it had taken so long for authorities to rescue him. He was trying to elude them, he said. The police report filled in the details:

> He remembers seeing all the lights, but told his father that he hid every time someone with a light went by, because he did not want to get in trouble for drinking.

The teenage girl who hosted the spontaneous party when her parents went out for the night initially told the police that Bennett was drunk when he arrived and that she had walked him part of the way home. Not until 5:00 a.m. did she admit the truth, that there had been more than a dozen people at the house, many of them drinking alcohol, all underage, and that she tried to clear them out around 11:30 before her parents returned home. Two girls said they were going to walk Bennett up the street, "but when they went outside with him and he was too drunk," they took him back inside and left him alone while they helped their friend clean up. That was the last time they saw Bennett.

Teen consumption of alcohol was only half the problem. The other half was the poor decision-making on the part of Bennett and his friends at the party, the lying that led to a delay before the police found Bennett, and even his panic at the thought of being caught by the police. All the teenagers involved exhibited a stunning lack of insight.

What scientists tell us is that insight depends on the ability to look outside oneself, and because that skill arises in the frontal and prefrontal lobes, it takes time to develop. The dynamic changes taking place in the brain are part of what make the adolescent years an age of exuberance. But a malleable, still-maturing teenage brain can be a scary proposition. Anything can happen—much of it not good. Teenagers may look like adults, they may even think like adults in many ways, and their ability to learn is staggering, but knowing what teenagers are unable to do—what their cognitive, emotional, and behavioral limitations are—is critically important.

4
Learning
A Job for the Teen Brain

"What did I do wrong?"

Often that's the second question I get from parents of teenagers. The first question is usually rhetorical:

"How could my [son/daughter] [fill in the blank]?"

Most of the parents who come up to me after a talk, or e-mail me or stop me at the grocery store, are exhausted or exasperated or both, and all of them could fill in the blank in that question with a whole host of perplexing actions, from "Why would my teenage daughter sneak out of the house in the middle of the night to be with her boyfriend after they just spent the whole weekend together?" to "How could my son raid the liquor cabinet of his friend's parents—and then leave the empty bottles behind to boot?!"

A neighbor of mine with a sixteen-year-old was flummoxed when she caught her son smoking pot in his room when he was supposed to be studying. That was bad enough, she told me, but what astonished her even more was the fact that he had the window wide open (it was the middle of winter, mind you) in order to air out his room—and the wind was blowing the smoke back *into* the room, under the door and down the stairs, where it wafted toward my horrified friend in the kitchen!

"How could he be that stupid?" she asked me.

Parents quickly blame themselves for a teen's poor behavior, even though they're not exactly sure how or why they're to blame. With biological parents, the guilt may come from passing on flawed DNA; and with biological and nonbiological parents or guardians, the guilt comes from questioning how they raised the child. In either case, you, the parent, are to blame, right? Yes, the two scenarios are different, but no, it's not because of the genes or anything you did or didn't do or because the teenager was somehow struck on the head and woke up as an alien species from the planet Adolescent.

Teenagers are different because of their brains and specifically because of two unusual aspects of their brains at this stage of their development. Their brains are both more powerful and more vulnerable than at virtually any other time in their lives. Even as they are learning things faster, their brains are eliminating gray matter and shedding neurons. How both of these facts can be true is because of something called neural plasticity.

Even as a teenager I used to wonder about the brain. Did it make a difference where a person grew up? How he or she grew up? Was the brain at all like the rest of the body—capable of changing depending on what went into it or what it was exposed to? I enjoyed turning these questions over in my head, and when I got to college they turned up again, only this time I began to have inklings of some of the answers.

During one summer while I was still in high school I volunteered at the Greenwich chapter of the Association for Retarded Citizens

(ARC), now known simply as the Arc, which aids people with intellectual and developmental disabilities. Some of those who regularly attended the Greenwich ARC were born with Down syndrome, and though they had varying abilities, most were self-sufficient. They were able to swim and to participate in the theater program; some even learned to read and write. Because of Greenwich's affluence, not only was the local ARC always well funded, but many of the kids came from very privileged backgrounds as well. To this day I remember being astonished when a limousine dropped off a tot for his day of activities with us. These children were really in an unusually enriched situation, and the effects of this gifted environment showed. Despite their handicaps and rather serious diagnoses, they were active and curious and engaged, and many were approaching milestones for reading and arithmetic close to those expected for normal kids their age. I knew that not only were they getting a great day at the ARC, but when they went home, they were often given physical therapy and tutoring there, too.

While at Smith, I had an opportunity to see what life was like for the mentally and developmentally disabled who did not have the same advantages as the children at the Greenwich ARC. I volunteered several hours a week at the Belchertown State School, a seventy-year-old state institution for the cognitively handicapped, located just a few miles from Smith. Belchertown's residents ranged from children to the very elderly, many of whom had spent most of their lives at the institution. Before it closed in 1992, Belchertown housed as many as 1,500 people, ages one to eighty-eight, living in thirteen dormitories. The hospital was understaffed, even after a local newspaper exposed overcrowding and maltreatment in the 1960s. When I volunteered in 1975, I primarily spent time in the children's dormitory. It was not a pleasant place. The rooms smelled of disinfectant, toys were few and far between, and many of the kids hadn't been bathed in quite some time. Like the children at Greenwich's ARC, some were more disabled than others, but even those who were more functioning seemed to lag far behind their peers at ARC. They sat in corners and rocked and had difficulty speaking, and their eyes appeared vacant.

This was a time at the height of the nature-versus-nurture debate, and my psychology and biology professors at Smith were keen on discussing how much a person's makeup, from personality to intelligence to likes and dislikes, is dependent on genes (nature) and how much on the influence of environment (nurture). There was clearly little nurturing going on at Belchertown, while at ARC there were always activities, directed therapies, teaching, and, most of all, stimulation.

At some point I realized the children at Belchertown who had the same disabilities and the same hurdles to overcome were far worse off than the kids at ARC in Greenwich, and at least from my limited viewpoint, environment seemed to be the overwhelming determining factor. It was pure and simple: the brains of the ARC children were being stimulated and encouraged, and the brains of the Belchertown children were not.

Like fingerprints, no two brains are identical. Everything we do, think, say, and feel influences the development of our most precious organ, and those developments trigger ever more changes until the thread of action and reaction is too complex to unwind or undo. Our brains, in essence, are self-built. They not only serve the particular needs and functions of the particular individual, but also are shaped—landscaped if you will—by the individual's particular experiences. In neuroscience, we refer to the human brain's unique ability to mold itself as plasticity. Thinking, planning, learning, acting—all influence the brain's physical structure and functional organization, according to the theory of neuroplasticity.

As far back as Socrates, some believed the brain could be "trained," or changed, much as a gymnast trains his or her body to balance on a high beam. In 1942 the British physiologist and Nobel Prize winner Charles Sherrington wrote that the human brain was like "an enchanted loom, where millions of flashing shuttles weave a dissolving pattern, always a meaningful pattern, though never an abiding one." In essence, the human brain, said Sherrington, was always in a state of flux.

Five years after Sherrington, Donald Hebb, a Canadian neuropsychologist, was struck by a kind of accidental inspiration that led to the first quasi-experimental test of the theory of brain plasticity. When the forty-three-year-old researcher took rat pups home from his lab at Canada's McGill University and gave them to his children as pets, he allowed the rodents to roam freely around the house. Hebb's inspiration was to compare the brains of these free-roaming pet rats with those of rats kept in cages in his lab. After several weeks he put both groups of rats through a kind of intelligence test involving a maze. The pet rats, which had free access to explore the environment of Hebb's home and unfettered interaction with one another as well as with Hebb and his family, performed significantly better on the maze test than the lab rats confined to small cages.

By the late 1990s researchers had confirmed a range of changes associated with experience and stimulation, including brain size, gray matter volume, neuron size, dendritic branching, and the number of synapses per neuron. The more stimulation and experience, they concluded, the larger the neurons, the bushier the dendrites, the higher the number of synapses, and the thicker the gray matter.

During my senior year at Smith College in 1977–78, I wrote my first professional journal article under the tutelage of Nico Spinelli, a professor in both the psychology department and the computer and information science department at the University of Massachusetts Amherst. He was doing pioneering experiments in the plasticity of the visual cortex. Previous research had looked at the brains of mammals raised in a deprived environment. Spinelli wanted to see if plasticity was still at work in a "normal" environment. So we took kittens raised with their mothers in a standard animal facility and gave them what's called avoidance training. In these experiments, a "safe" and an "unsafe" stimulus were associated with two different visual stimuli: vertical lines and horizontal lines. As the kittens learned to associate the safe stimulus with either the horizontal or the vertical lines, the number of neurons in those parts of the visual cortex expanded. The results, which were published in the journal *Science*, confirmed "that early learning produces plastic changes in the structure of the developing brain," or, to put it more simply, young brains are shaped by experience.

Of course, adult brains can be shaped by experience as well. Researchers in neural plasticity have found that even in the last decades of life, adult brains can be remodeled, just not as easily or as constantly as during childhood and adolescence. Whereas kids' brains will respond and change in response to virtually any stimulation, so-called adult plasticity occurs only in specific behavioral contexts. For instance, cab drivers in London (a notoriously difficult city to navigate) have been found by scientists to have an enlarged hippocampus particularly in the area responsible for spatial memory. Violinists and cellists, who must use their hands fluidly and rapidly, have been shown to have an enhanced motor cortex. And in an unusual experiment conducted several years ago, Patricia McKinley of McGill University was able to show that learning the tango, which involves both complex movement and a fine sense of balance, improved the ability of senior citizens, ages sixty-eight to ninety-one, to switch between two different cognitive tasks. "Plasticity," then, is just another way of saying "learning."

In the first few years of childhood there is a critical period of plasticity in which learning comes quickly and easily. Evolution experts believe this is the brain's way of helping us adapt early to the specific environment in which we are raised. The concept is the same as that of imprinting, whereby a baby duckling develops a keen and powerful preference to follow the mother duck over any other. When I was five years old, I saw this in action, although I obviously didn't know it at the time. It was Easter, and my baby brother had just been born. Perhaps because of that, friends of my parents gave me my own "baby"—a baby chick, that is, much to my parents' consternation. I loved that fuzzy little animal and was absolutely fascinated that it would follow me around the house, through the swinging door between the kitchen and the dining room, even out of the house and around the yard. Because I was with the chick almost from its birth, it had determined I was its mother. Years later I would read the children's book *Are You My Mother?* by P. D. Eastman to my sons. Basically, the book is really all about imprinting. A young hatchling leaves its nest too early while its mother is out foraging for food, and goes on a journey, asking every animal and object it meets—a kitten, a hen, a dog, a cow, a car, even an enormous power shovel—the question of the title. Luckily the power shovel lifts the young bird up and deposits it back in its nest beside its real mother.

Five-year-old me, of course, was the only mother my baby chick had. Unfortunately, the end of the relationship was sudden and brutal. About a week after Easter, after I'd just gotten home from kindergarten, my baby chick was once again following me all over the house, but this time, as I skipped between the kitchen and the dining room, the little hatchling failed to make it through the swinging door and was squished. I cried for days.

Thirteen years later, as a freshman at Smith, I created my own chick-imprinting experiment for a class in advanced biology. In order to imprint them to sound, I exposed my baby chicks to a specific sound or tone every day over a week. At the end of this training period, the chicks were placed on a kind of runway and were then exposed to two sounds, one of them being the familiar tone I'd played for them for seven straight days. Every one of the chicks toddled toward the familiar tone: they had imprinted to sound. I remember this so well because my mother was visiting me at the time of the experiment and she helped me type the results!

But how does learning actually happen? Young brains and old brains work much the same way, by receiving information from the senses—hearing, seeing, tasting, touching, smelling. Sensory information is transmitted by synapses through a network of neurons and is stored, temporarily, in short-term memory. This short-term memory region is highly volatile and is constantly receiving input from the nearly continuous information our senses encounter every minute of our waking life. After information is processed in the short-term memory region, it is compared with existing memories, and if the information matches, it is discarded as redundant. (Brain space is too limited and too precious to allow duplicates to take up neural real estate.) If, on the other hand, the information is new, then it is farmed out to one of several locations in the brain that store long-term memories. Although nearly instantaneous, the transmission of sensory information is not perfect. In the same way that the otherwise seamless signal coming from your TV is occasionally interrupted, briefly distorting the picture, so, too, does degradation occur as information races up and down the axons of your brain's neurons. This explains why our memories are never perfect, but have holes or discontinuities, which we occasionally fill in, albeit unconsciously, with false information.

The brain is programmed to pay special attention to the acquisition of novel information, which is what learning really is. The more activity or excitation between a specific set of neurons, the stronger the synapse. Thus, brain growth is a result of activity. In fact, the young brain has more excitatory synapses than inhibitory synapses.

The more a piece of information is repeated or relearned, the stronger the neurons become, and the connection becomes like a well-worn path through the woods. "Frequency" and "recency" are the key words here—the more frequently and the more recently we learn something and then recall it or use it again, the more entrenched the knowledge, whether it's remembering the route between home and work or how to add a contact to your smartphone's directory. In both cases, the mental machinery of learning is dependent on the synapse, that minuscule space where packets of information are passed from one neuron to another by electrical or chemical messengers. For these neural connections to be made, both sides of a synapse need to be "on," that is, in a state of excitation. When an excitatory input exceeds a certain level, the receiving neuron fires and begins the molecular process, called long-term potentiation, by which synapses

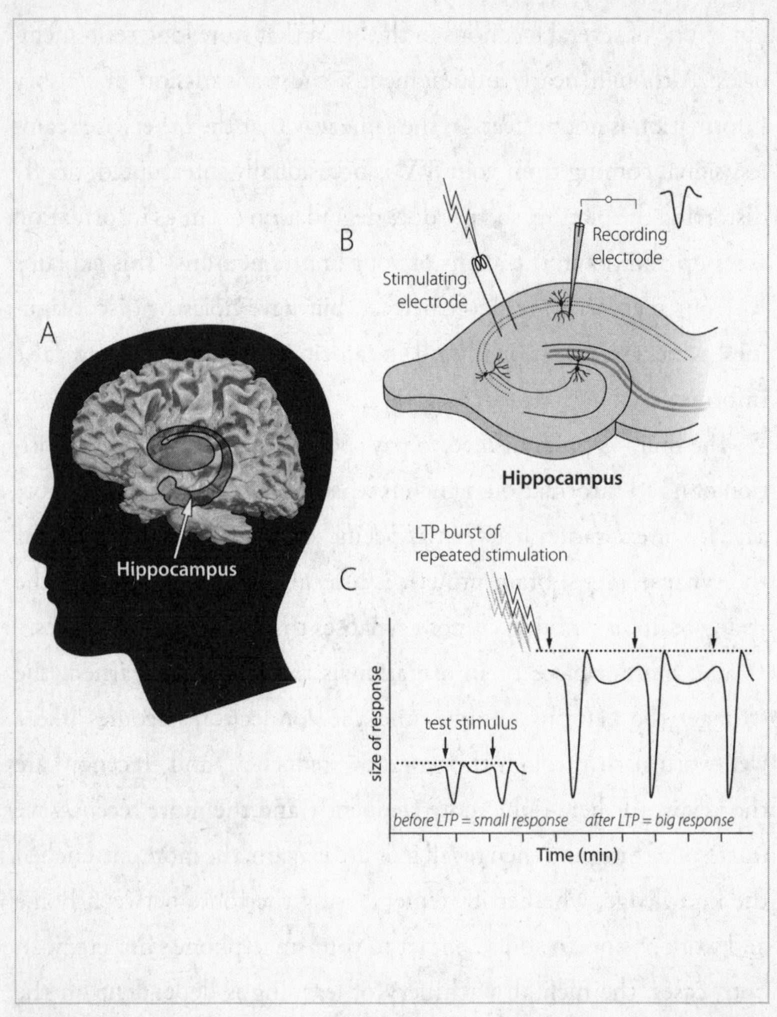

FIGURE 10. Long-Term Potentiation (LTP) Is a Widely Used Model of the "Practice Effect" of Learning and Memory: A. The hippocampus is located inside the temporal lobe. B. Brain cell activity recorded in hippocampal slices from rodents shows changes in cell signals after a burst of stimulation. C. LTP experiments commonly record repeated small responses to stimuli until a burst is given (akin to the "practice effect"), after which point responses from the neuron to the original stimulus become much larger, as if "memorized" or "practiced."

course, the quickest route down becomes worn by use. Ruts develop. By the time the last competitors race through the gates, the route is so deeply entrenched in the snow that they can't ski out of it, nor do they need or want to. The deeply imprinted line, in fact, guides them down without their having to search for it.

The process of fine-tuning and turning off neuronal connections that may have been made in childhood but are no longer needed is called pruning, and it accelerates during mid- to late adolescence when unneeded synapses are removed. Scientists call this pruning phase a kind of "neural Darwinism," in which only the "fittest," that is, the most used neurons, survive. What accounts for gray matter loss so early in a person's development when many cognitive and behavioral functions have yet to fully develop? Researchers have discovered in the past few years a direct correlation between an adolescent's decrease in gray matter and the increase in white matter. Scientists know that gray matter continues to decrease through adulthood, especially after the age of sixty, but they also believe that gray matter loss in adolescence is a distinctly different process. In later life gray matter declines as a function of degenerative processes, that is, cell shrinkage and death, whereas in adolescence gray matter decline is a product of the brain's plasticity. ("Use it or lose it.")

Researchers at UCLA discovered not only that effective pruning increased brain efficiency but also that higher intelligence could be correlated with prolonged, accelerated neuronal growth in childhood followed by vigorous cortical pruning in adolescence. What is less, it turns out, is actually more, and this is why in the midst of all the chaos of the teenage years, adolescents are developing a leaner, more efficient adult mental "machine." Adults have an advantage over children and teens in that their white matter is more extensive, meaning

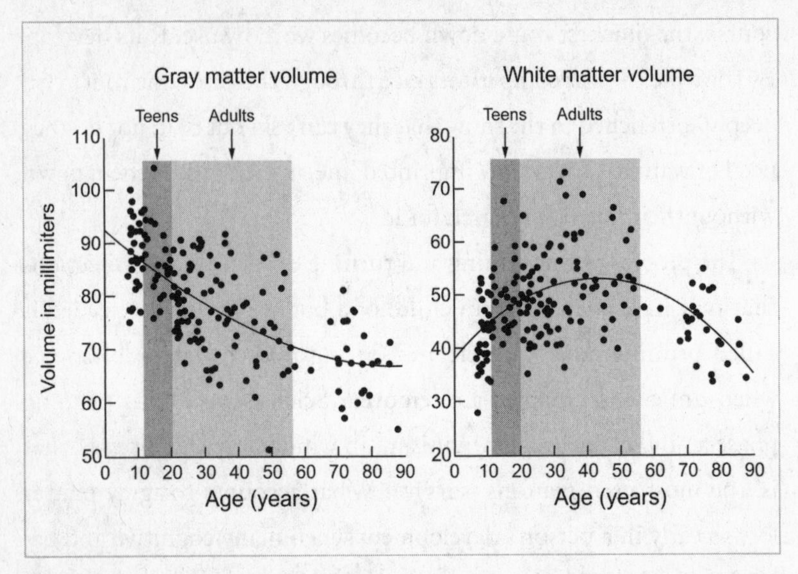

FIGURE 12. Gray Matter and White Matter Develop Differently Throughout Life: Children and teens have more gray matter and synapses than adults, because as we age our brains remove unnecessary connections for greater efficiency. However, in old age white matter also diminishes, contributing to age-related cognitive issues such as memory loss and abnormal conditions such as dementia.

that they have more speedy connections between brain areas, such as the frontal lobes.

The real news in all of this science is that the teen brain is more than capable of learning, and this fact should not be taken for granted! LTP is indeed more robust in teens. Adolescent animals similarly show faster learning curves than adults, and scientists wanted to know whether this was because they had better synaptic plasticity, given all the increase in excitatory synapses, which are required for LTP. Studies were performed on brain slices from rats to look at the LTP in an adolescent rat versus an adult rat. They found that the strength of LTP was "way better" in the adolescents. The "before" and "after" comparisons following the burst of stimulation showed that synapses in adolescent slices had about one and a half times as much increase and that it lasted a lot longer.

So what this means is that memories are easier to make and last longer when acquired in teen years compared with adult years. This is a fact that should not be ignored! This is the time to identify strengths and invest in emerging talents. It's also the time when you can get the best results from remediation, special help, for learning and emotional issues. We've all long thought that the IQ with which you were "branded" in grade school after taking one of those aptitude tests was the final word on your intellectual destiny. Not true. There is solid data to show that your IQ can change during your teen years, more

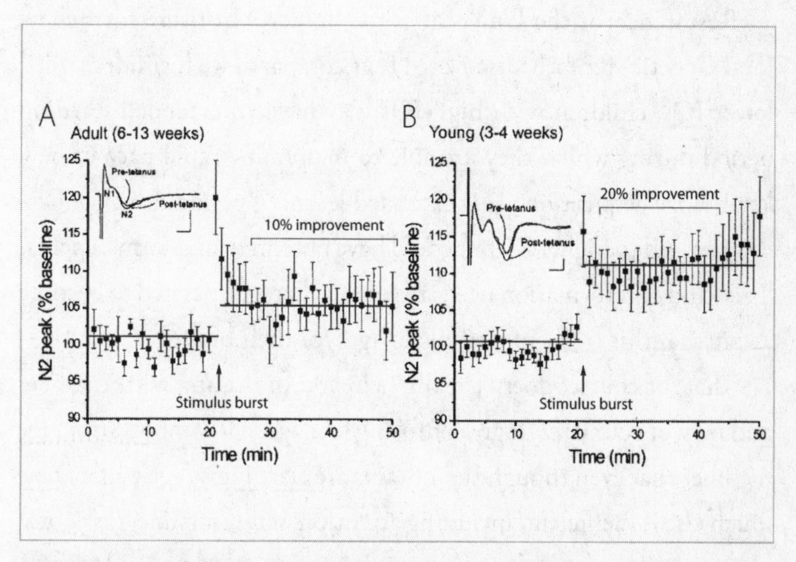

FIGURE 13. Adolescents' Synaptic Plasticity Is "Way Better" Than Adults': To test if teens have better learning abilities than adults because they have better LTP, researchers compared adolescent hippocampal brain slices with those from adults. The signal after the burst stimulation in the adolescent (B) was much higher, and lasted longer, than in the adult (A).

than anyone had ever expected. Between thirteen and seventeen years of age, one-third of people stay the same, one-third of people decrease their IQ, and a remarkable one-third of people actually significantly raise their IQ. The changes in increased IQ were also associated with changes in brain scans. When verbal IQ increased, so did gray matter in the center of the brain, responsible for articulating speech. When nonverbal IQ increased, so did gray matter in the area of the brain related to hand movements. The frustrating part of this study is that the researchers did not track what these people did during those formative years. We'd love to know the secret of how to make your IQ go up during your teen years. At least we are already learning the kinds of things that can make it go down, but we will come back to this later.

Researchers at the University of Colorado's Institute for Behavioral Genetics recently discovered that compared with children with lower IQs, children with high IQs may have an extended learning period during which they are able to maintain a rapid pace of new knowledge acquisition. This extended learning period does not necessarily lead to higher IQ but could have potential long-term benefits. This kind of information needs to get out there: teens need to become aware that this is one of the golden ages for their brains!

That, of course, doesn't really help you in dealing with the here and now of your dizzyingly confusing teenagers. It's important to remember that even though their brains are learning at peak efficiency, much else is inefficient, including attention, self-discipline, task completion, and emotions. So the mantra "one thing at a time" is useful to repeat to yourself. Try not to overwhelm your teenagers with instructions. Remember, although they look as though they can multitask, in truth they're not very good at it. Even just encouraging them to stop and think about what they need to do and when they need to do it will help increase blood flow to the areas of the brain involved in multitasking and slowly strengthen them. This goes for giving instructions and directions, too. Write them down for your teens in addition to giving them orally, and limit the instructions to one or two points, not three, four, or five. You can also help your teenagers better manage time and organize tasks by giving them calendars and suggesting they write down their daily schedules. By doing so on a regular basis, they train their own brains.

Perhaps most important of all, set limits—with everything. This is what their overexuberant brains can't do for themselves. So be clear about the amount of time you will allow your teenager to socialize "virtually," either on the Internet or through texting. Best-case scenario: limit the digital socializing to just one to two hours a day. And if your teenager fails to comply, take away the phone or the iPod, or limit computer use to homework. Also, insist on knowing the user names and passwords for all their accounts.

None of this means your kids are going to immediately go along with the program. In fact, it's virtually a certainty that there will at least be occasional slip-ups, perhaps a lot of them. That's why it's up to you to keep tabs, to check on teenagers as they do their homework and spend time on the computer. The more on top of it you are, the fewer the temptations for your adolescents, and the fewer the temptations, the more their brains will learn how to do *without* the constant distractions.

In dealing with those unexpected emotional outbursts from your teenage sons and daughters, something as simple as counting to ten before responding can help you stay calm. Being angry, or treating the meltdown like a childish tantrum, is not advisable. Adolescents believe they're adults, and though we know better, the more you treat them that way, the greater the chance they'll actually try to act that way, too. Because I'm a doctor and a scientist, I could sit my kids down and tell them, Look, you don't believe me when I say that you're being irrational or impulsive or overly sensitive, but let me tell you why it's your brain's "fault." By the time you finish reading this book, you'll be able to do the same thing. And trust me, it works. I've seen it not only with my own sons but also in talking to teenagers after I've given presentations in high schools. They're actually fascinated by the neuroscience and that there's a logic, or rationale, behind what otherwise seem like inexplicable upheavals in their lives. You do run the risk, however, that these "brain explanations" become a kind of adolescent ammunition.

"My brain made me do it." Your teenage son might be tempted to say when he decides to drive off with Dad's car and not tell anyone until he comes homes after midnight.

"Well, no," you have to say, "your brain is sometimes an explanation; it's never an excuse."

Your teenagers are knowledgeable and self-aware enough to know that they are not automatons, and this means they have the capacity to modify and the responsibility for modifying their own behavior. This is what you must remind them—and then remind them again and again and again. The brain science isn't an excuse for crazy, stupid, illegal, or immoral behavior. It's an explanation and a framework, about which they should be encouraged to read more. In the same way that I called my sons after I heard the news of Dan's drowning, when you read or hear about something similar, call your teenagers, or sit down with them, and remind them why these things happen. You can't expect them to automatically grasp innuendo and more subtle connections, so the better thing to do is err on the side of overkill and point out the obvious, explicitly. I did this so many times with my kids that they named me Captain Obvious!

There is a logical reason why plasticity is front-loaded in childhood and adolescence: survival depends on knowledge of one's environment, so the young brain must be flexible and moldable depending on the type of environment in which the person is growing up. The growth of synapses makes teens sensation-seeking learning machines, but the fact that their brain signals can also easily slide off the tracks makes those growth spurts somewhat dangerous. Evolutionarily, being open to new ideas, having new things to learn, leads to useful experiences that are necessary for survival.

For adults, it's our myelin that in part allows certain brain signals to speed their way to the frontal lobe, where we respond by putting the brakes on that impulse to skydive or drive at 120 miles an hour. So pruning happens just as myelin growth increases, giving adolescents a short window to experience the world and figure out what will make them happier, healthier, and, one can only hope, wiser. This delicate balance is why some teen behaviors can appear so confounding, like the teenage boy a colleague told me about who was stopped on a side street for driving at 113 miles an hour and given a speeding ticket. The boy was furious—not that he thought he shouldn't have been ticketed, since he did admit he was driving way over the speed limit, but because the violation was for "reckless driving," when in fact, he told his father, he'd thought about it and planned it all out well in advance. He knew exactly what he was going to do and where to do it, and even picked a straight road with little traffic and good weather to do it in. On the other hand, he then went out and drove 113 miles an hour!

There's an additional, evolution-based reason for this apparent conflict of self-interest. Scientists at University College London recently asked fifty-nine young people, ages nine to twenty-six, to guess the odds that certain bad things might happen to them. The forty unfortunate events ranged from getting lice to getting seriously injured in an auto accident. After the subjects made their guesses, they were told the real odds of those bad things happening. Then they were asked to project the odds again. The majority of the subjects were good at remembering the actual risk if it wasn't as bad as their original guess. But the adolescents were worse at recalling the risk when the risk was *worse* than their original guess. As it turns out, there are more areas of the brain that process positive information, whereas negative information is centered in the prefrontal cortex. In other words, adolescents have less ability to process negative information than adults do, and so they are less inclined *not* to do something risky, and less likely to learn from the ensuing mistake or misadventure, than adults are.

Once someone is out of adolescence, synaptic plasticity, and learning, require more effort. In the same way that a young adult eventually settles down into a life and a routine, so, too, does the brain. The middle-aged man who played electric guitar in a rock band in high school would have trouble picking up the instrument again at age forty-five. Twenty-five years ago, those guitar-playing neurons were in constant use, but by the time he reaches adulthood they've been dormant for so long they were essentially left behind, just like that electric guitar in the attic. Adults also have less glutamate and dopamine and fewer receptors available; therefore they are less cognitively flexible.

But don't tell my dad that. He's now in his early nineties and still active, and his favorite gadget is his iPad, which he never puts down. He's always e-mailing me, sending me virtual clippings from medical articles he's read, attaching them to an e-mail that just reads, "I thought you'd like to see this." My father's brain is engaged all the time, and the Internet has helped him stay stimulated and on top of current issues and events. If my dad had been born just twenty years earlier than he was, I'm not sure what he'd be doing now or what he'd be able to do. My mother, who is also in her nineties, prefers to play solitaire on the iPad. She says it's easier than shuffling cards. Having grown up in England, she worked in British Intelligence during World War II, and her brain remains sharp. The good news about brain plasticity is that it may peak in childhood and adolescence but it never entirely stops—at least not until we do. The more you learn, the easier it is to learn the next thing.

5 Sleep

"He's lazy."

"She deliberately disobeys me."

"He just wants to sleep away the day."

One of the chief complaints I hear from parents of teenagers is their frustration at not being able to get their kids to go to bed at a decent hour every night, and then not being able to coax them out of bed in the morning. I've heard stories of parents who have tried cajoling, scolding, threatening. They've torn the covers off their teens' beds and banged on pots and pans, all to no avail. One mother would try to wake her son every fifteen minutes until he finally got out of bed, but no matter how early she started the process, he was still always late for school. She was a nervous wreck every morning because she also had to get to her job. One day her son delayed getting out of bed for the umpteenth time and she had to drive him to school. On the way, he fell asleep in the car and then refused to get out and go to class! Completely worn out, she finally drove to work—*she* wasn't going to be late—and left her son in the car, asleep. At lunchtime, she went out to the parking lot. This time he not only was angry at her for waking him up but also said he was hungry!

A boy who sleeps *that* much needs a physiological workup to find out what's causing the extreme fatigue. That is an extreme case, of course, and that particular teenager may well have a physiological problem that exacerbates his bodily tiredness. I tell you this story, however, to emphasize that teenagers who refuse to go to bed at night or get out of bed in the morning are not slothful, nor do they lack discipline, and refusing to heed your pleas to wake up is also not a sign of rebelliousness—we'll get to plenty of those signs later, including the evolutionary explanation for teenage rebellion. I tell you this story because the infuriating behavior of teens when it comes to sleep is actually completely normal.

Let me explain.

Sleep is one of the most important aspects of daily life, and yet it is also one of the least understood. What we do know about sleep is that it is critical to the health of every human being. Sleep patterns, or chronotypes, change across the life cycle and in the same way in all species. Infants and children are "larks"; that is, they wake up early and go to sleep early. Adolescents are "owls," waking late and staying up until the wee hours of the morning. The technical terms for "larks" and "owls" are early and late sleep chronotypes. Sleep patterns are controlled by a complex web of brain signaling and hormones, both of which are regulated by maturational stages. In most species, this temporary shift to late-night wakefulness during adolescence reverts more to the "early to bed, early to rise" pattern in adulthood.

Teenagers can be, and are, forced to abide by the adult chronotype, with early rising for school. However, this early rising does not result in an early bedtime: the teen brain doesn't adjust at the

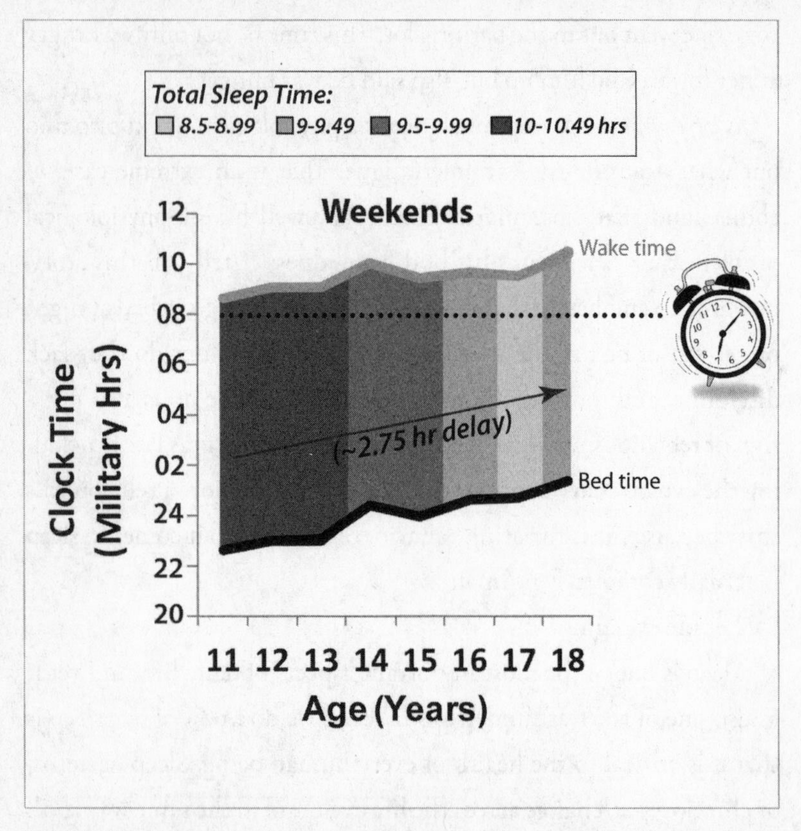

FIGURE 14. The Developmental Control of the Circadian System: Teens have a tendency to stay awake longer and sleep later. This graph compares teens' sleep hours on the weekends, when study subjects slept for the amount of time their bodies wanted, with those during the week when they were "artificially" woken up for school by an alarm clock.

other end of the day, and instead has a tendency to hold on to that part of its pattern. The result is a shrunken sleep period. However, on weekends, one sees teenagers immediately slip back into late-morning awakenings, as their internal clock prefers. If they are allowed to sleep as long as they like, teenagers will get around 9 to 10 hours of sleep per night. But if they are made to wake up for school, as shown in

Figure 14, they are chronically losing 2.75 hours of sleep daily. This is thought to contribute to a chronic sleep deprivation syndrome.

Because so much is going on in adolescents' brains, and they are learning so much and at such a fast pace, teenagers need more sleep than either their parents or their much younger siblings. In an earlier chapter I told you about the pruning that takes place in the teenage brain during puberty. When do you think that actually takes place? Yep, that's right, when they're asleep. Bedtime isn't simply a way for the body to relax and recoup after a hard day of working, studying, or playing. It's the glue that allows us not only to recollect our experiences but also to remember everything we've learned that day. Sleep isn't a luxury. Memory and learning are thought to be consolidated during sleep, so it's a requirement for adolescents and as vital to their health as the air they breathe and the food they eat. In fact, sleep helps teens eat better. It also allows them to manage stress.

Scientists have calculated that the average adolescent actually requires nine and a quarter hours of sleep. (The Centers for Disease Control and Prevention recommends adolescents get eight and a half hours to nine and a half hours of sleep a night.) Only about 15 percent of all American teenagers actually get that much on a regular basis. Worse, most American teens sleep fewer than six and a half hours a night. How does this happen? Beginning at around ages ten to twelve, young people's biological clock shifts forward, revving them up by about seven and eight o'clock at night and creating a "no sleep" zone around nine or ten o'clock at night, just when parents are starting to feel drowsy. One reason is that melatonin, a hormone critical to inducing sleep, is released two hours later at night in a teenager's brain than it is in an adult's. It also stays in the teenager's system longer, and this is why it's so hard to wake your high schooler up in the morning. Adults, on the other hand, have almost no melatonin in their system when they wake up and therefore don't have the same groggy feeling.

One of the unfortunate consequences of making kids wake early is that it does in fact put the squeeze on their sleep time. There are also so many more distractions at night that can keep a teenager up way past bedtime. In my generation, staying up late meant sneaking a book and a flashlight under the covers. Today's generation can feed its insomnia in any number of electronic ways, especially texting, making winding down for an already unnaturally early sleep time that much more difficult. This may be why so many of us with teenagers turn in well before our kids do, but always with a sense of worry about what can happen when we close our eyes and are no longer "on watch."

Awake/rest states, like sleep, have also been shown to strengthen learning. In laboratory studies, when rats are given a maze to explore, their brains show an expected increase in activity. When they are split into two groups, with one group of rats getting a rest after their explorations and the other none, those given the downtime remembered the maze significantly longer. This was shown in a study in Boston where two groups of learners were compared. The first group was given a task to learn early in the day for several days in a row. After each training session, people showed an improvement in the task, called a practice effect. However, when they returned the next morning for another practice/learning session, they showed a slight decay in their skill compared with immediately after the end of their previous session. We all know this phenomenon: a tennis or golf lesson leaves you feeling great, but then if you go out the next day, you are never as good as at the end of the earlier lesson—frustrating! The

Boston researchers wanted to know if sleep could help. So they repeated the learning sessions with another group of learners, and instead of doing the practice/learning sessions early in the day, they did them just before the people went to bed at night. They discovered an astonishing effect: when the subjects returned the following evening for their practice session, they did not show any decay whatsoever and were able to start additional learning right where they left off the day before! So there was no "two steps forward, one step backward" effect. These human studies substantiate a large amount of animal data showing how LTP can be affected by sleep deprivation. Brain slices from sleep-deprived rats, even after one day of sleep deprivation, show a diminished capacity for LTP compared with data from well-rested rats, and after two days of sleep there is even *more* impairment.

More recently, researchers at Brown have been looking at the effect of sleep on motor learning that accompanies piano lessons and comparing brain scans of people who learned finger movements right before sleep versus those who learned these movements but did not go to sleep immediately after their lesson. The Brown scientists were able to show that learners who "slept on it" showed better accuracy than those who did not, and this translated into more measurable activity during slow-wave sleep in the supplementary parts of the motor cortex responsible for coordinating patterned activity. Yuka Sasaki, an author of the study, concluded, "Sleep is not just a waste of time." Couldn't have said it better myself.

It is not just sleep that helps learning; even simply being in a restful state helps. At the University of Michigan researchers asked students to perform basic cognitive tests in order to fatigue their brains. They were then assigned to take either a fifty-minute walk in an arboretum or a fifty-minute walk into downtown Ann Arbor. The downtown walk was mostly on a heavily trafficked street. When both groups of students were retested after their walks, the performance of those who took a walk in nature was significantly better than the performance of those who had been assigned to walk into the busiest part of town. A week later the results held, even though the conditions for the two groups were reversed. In other words, those who previously had been told to walk into Ann Arbor now were told to stroll through the arboretum, and vice versa. Again, the group that took a restful walk through the garden outperformed the group assigned to walk into town. What the scientists determined was that the busy urban environment made more demands on the students' directed, or voluntary, attention, which taxes the brain. The natural setting of the arboretum, however, allowed students to rest their directed attention and let their minds wander. Downtime, whether it is a good night's sleep, a nap, or simply a few quiet moments of relaxation in the middle of the day, is important for turning learning into long-term memories.

In experiments done on high school students at Harvard Medical School and Trent University in Canada, it was discovered that consolidation of memories happens in two stages during sleep: slow-wave sleep and rapid eye movement (REM) sleep. Early in the sleep cycle of the teenager, the brain enters a slow-wave stage, which is the deepest sleep state. As a child goes through puberty, this deep slow-wave sleep decreases by as much as 40 percent. During REM sleep, which happens later in the sleep cycle, the brain puts on a kind of show, reenacting through dreams the information learned and further solidifying the information for storage in the brain's memory areas. This is why it is so important for teens to get *more* than just a good night's sleep before an exam. They need to get that good night's sleep right *after* studying for the exam.

My son Will, who has always been prone to asking "Why?" about everything I tell him or ask him to do, now finally agrees. When he was in high school, he was tempted to pull an all-nighter before a test. I informed him it would be better to study for a little while and then get a good night's sleep. When he asked why, I told him about the sleep-wake cycle and how it differs in teenagers. He took my advice, studied for a bit, and then went to bed. He may have woken up early to do some last-minute reviewing as well. The next day, when he came home from school, he happily announced I was right. He said he not only felt confident he'd done well on the exam, but also seemed to know the material better in the morning than the night before—all because he'd given his brain time to convert what he'd learned into memories while he was sleeping.

Sleep, however, doesn't just strengthen learning and memories. Researchers at the University of Notre Dame and Boston College collaborated recently on a memory study and found that sleep not only consolidates memories but also prioritizes them by stripping them down into their components and then organizing those components according to their emotional importance. So, for instance, when subjects were shown a photograph of a tiger in a wooded landscape before they went to sleep, they remembered the tiger better than the trees in the background. Evolutionarily, the ability to remember the most emotional part of a real event makes sense, especially when that emotion is fear and it's the adrenaline rush that gets you out of harm's way as fast and as far as your legs can take you.

When I've given talks to teenagers and I tell them about what an exciting time this is for their brains and how easy it is for them to learn new things—especially if they sleep on that knowledge—there are always a few smart alecks who say, "Cool, that means I don't have to start studying until right before I go to bed." I have to tell them, "No, just before you go to bed shouldn't be the first time you see the information. Your brain isn't *that* responsive. This is just a good time to review."

Research over the past decade has confirmed the relationship between sleep and learning in adolescents. In one study, moving the start time for school back just seventy minutes, from 7:30 to 8:40 in the morning, had a statistically significant effect on the grades of seven thousand high school students in Minneapolis and Edina, Minnesota. Compared with students in schools that maintained the earlier start time for the school day, the students in the districts whose schools started later reported they got more sleep, earned better grades, and experienced fewer episodes of depression. When high schools in Jessamine County, Kentucky, pushed back the first bell by an hour, attendance rose along with standardized test scores, and when high schools in Fayette County, Kentucky, did the same thing, the number of students involved in car crashes dramatically decreased—while rising in the rest of the state! At Concord Academy, where my older son attended high school, my "Teen Brain 101" lecture helped convince administrators to at least switch its exam schedule, so that instead of being given at 8 a.m., tests were given at 10 a.m. The students, I'm told, scored better with the later start, and the school has maintained its later exam schedule. For once I had managed to be a hero for my son rather than an embarrassment!

Starting the school day later seems a natural next step, but even with the new scientific findings the large majority of schools across the country have not adjusted their start times. The reason given by most boards of education is that starting the school day later in the morning would cause disruption to after-school activities and inconvenience both teachers and parents. However, according to the University of Minnesota's Center for Applied Research and Educational Improvement, when those Edina and Minneapolis high schools adjusted the start time of their day, after-school jobs and activities were not severely affected. Scheduling was slightly more complicated but not disruptive, and participation, for the most part, remained about the same. Some schools even reported that the more rested athletes performed better on the playing fields.

A few years ago, scientists at Washington University in St. Louis studied the sleep-learning connection from the opposite direction. That is, how does learning affect the need for sleep? In fruit fly experiments—fruit flies actually have sleep-wake cycles similar to those of humans—the researchers looked at how young fruit flies responded to being raised in an enhanced social environment. After flying around in a large, well-lit chamber with other young fruit flies, they all grew more branches on their neurons with many more synapses. They also required two to three more hours of sleep than fruit flies raised in isolation. But what also surprised the researchers was that after sleep, the synapses in these social fruit flies that had been given space in which to roam returned to normal size. Of the twenty thousand cells in a single fruit fly's brain, only sixteen neurons were needed to consolidate the day's learning into memories. The brains of those flies that were denied sleep after being exposed to the enriched environment continued to have synapses that were larger and denser. In other words, learning appeared to be related to the pruning of synapses during sleep, leaving space for new ones to grow. Sleep, it seems, provides time—and preserves energy—for the brain to pick out the most salient information from the day's activities and consolidate that information into memories, discarding the rest. Like almost everything else about human life, the brain is a finite organ, with a finite amount of space. It makes sense that if the brain simply kept adding synapses, it would soon reach a limit and all learning would cease. The more you learn, the more you need to sleep, it would seem.

So what happens when teenagers *don't* get enough sleep? Nothing good, that's for sure. Sleep deprivation inhibits the necessary synaptic pruning or prioritizing of information. And a lack of good sleep habits results in much more than a tired body and mind. It can have profound and lasting effects on teenagers and could contribute to everything from juvenile delinquency to depression, obesity, high blood pressure, and cardiovascular disease. Studies have shown that teenagers who report sleep disturbances have more often consumed soft drinks, fried food, sweets, and caffeine. They also report less physical activity and more time in front of TV and computers. Another study found that teenagers who had trouble sleeping at ages twelve to fourteen were two and a half times more likely to report suicidal thoughts at ages fifteen to seventeen than adolescents with good sleep habits.

Japanese researchers have found that teens who used their cell phones after "lights out" not only had reduced time asleep but also were at increased risk of mental health disorders, including self-harm and suicide. Colleen Carney at the National Institute of Mental Health (NIMH) has been showing that insomnia can worsen depression, and that behavioral therapy (not sleep medications) aimed at better sleep habits can help lower depression rates. Scientists don't yet fully understand the connection between shortened sleep time and poor mental health in teenagers, but the fact that teens spend more time on their phones than most adults seems indisputable. Talking, of course, is only one use of a cell phone. More than five billion text messages are sent every day in the United States. Not surprisingly, a large portion of the texting population is teenagers. According to one recent study, each teen sends an average of 3,300 texts every month. (Girls average more: 4,050 texts a month.) Researchers at a sleep disorders clinic at JFK Medical Center in New Jersey estimate that one in five teenagers actually interrupts his or her sleep in order to text. The participants in their study, all of whom had come to the clinic for sleep issues, reported sending and receiving an average of thirty-four texts every night—after going to bed! The texts were sent and received from ten minutes to four hours after these teens went to bed, and the adolescents were awakened by a text message at least once a night. There was a slight gender gap in the research. Girls were more likely to text after going to bed, whereas boys were more likely to be awake, playing games on their cell phones. (Excessive or obsessive texting is also now being treated like an addiction.)

Poor sleep habits may even have a role in juvenile delinquency. The *Journal of Youth and Adolescence* reported in 2012 that teens who slept seven hours or fewer a night had a significantly higher rate of property crimes such as shoplifting, vandalism, and breaking and entering than peers who had eight to ten hours of sleep a night. Those teens who slept five or fewer hours a night had significantly higher rates of violent crimes, including physical fights and threatened violence using a weapon, compared with teens who slept eight to ten hours a night. The relationship is not yet fully clear since stressful environments have an effect all their own on behavior and can also affect sleep. In 2011 a large study undertaken by the Centers for Disease Control and Prevention found a correlation between poor sleep habits in teens and increased risk of unhealthy habits, including use of cigarettes, alcohol, and marijuana. Researchers in Italy have reported similar findings. There is not a single part of a teenager's life that is not adversely affected by a lack of sleep.

Physiologically, poor sleep can result in:

- Skin conditions that worsen with stress, like acne or psoriasis
- Eating too much or eating the wrong foods
- Injuries during sports activities
- Rise in blood pressure
- Susceptibility to serious illnesses

Emotionally, bad sleep can make teenagers:

- Aggressive
- Impatient
- Impulsive and inappropriate
- Prone to low self-esteem
- Liable to mood swings

Cognitively, poor sleep can cause:

- Impairment of the ability to learn
- Inhibition of creativity
- Slowing of problem-solving skills
- Increasing forgetfulness

One unfortunate consequence of sleep deprivation in everyone, but especially in teenagers, is that increasingly insomniacs are turning to artificial stimulants to keep themselves awake during the day. Some of those stimulants, such as Ritalin and Adderall, which are normally given for attention deficit hyperactivity disorder, are illicit, but the most popular and completely legal ones are energy drinks. Just the names of these drinks can give a parent pause, and it's no wonder they appeal to the thrill-seeking adolescent: Red Bull, Full Throttle, CHARGE!, NeuroGasm, Hardcore Energize Bullet, Eruption, Crave, Crunk, DynaPep, Rage Inferno, SLAP, and my personal favorite (the name, not the drink!): Venom Death Adder. While the US Food and Drug Administration restricts the amount of caffeine in soft drinks to a maximum of seventy-one milligrams of caffeine per twelve-ounce can, it does not restrict the amount of caffeine in energy drinks, because they are classified as dietary supplements. The amount of caffeine in an energy drink varies from eighty milligrams to as much as five hundred milligrams. Teens and twenty-somethings also sometimes mix the high-caffeine energy drinks with alcohol so that they don't feel inebriated as long as they're wired on caffeine. The problem with mixing the Red Bull–type energy drinks of the world with booze is that instead of passing out, a seriously inebriated person is still up and around and likely to be operating under the false assumption of being able to perform complex tasks, such as driving a car, when he or she is not.

Some surveys suggest that anywhere between 30 and 50 percent of all adolescents and young adults consume energy drinks; this figure perhaps explains the exponential rise in caffeine overdoses being seen in hospital emergency rooms. In 2013 the US government's Substance Abuse and Mental Health Services Administration reported energy-drink-related ER visits increased tenfold between

2005 and 2011, from fewer than two thousand to more than twenty thousand. Some studies report that the average high school student drinks as many as five cans of energy drinks a day to compensate for a lack of sleep.

As critical a role as sleep plays in the learning process of teenagers, so do parents and guardians, and there are things you can do to encourage your teenage sons and daughters to get enough sleep, beginning with taking the TV and computer out of the bedroom. Because they're in a chronic state of sleep deprivation, you have to help them get their homework done and get them to bed early. When they come home from school, ask them how much homework they have, try to get an inventory of what they have to do, and help them prioritize. If an assignment includes something creative, suggest it be done first because it involves more complex cognitive skills and more focus. Check on them over the course of the evening, but try not to do it judgmentally. The worst thing to do at 9:30 at night when you find out the English lit essay still hasn't been written is to shout or scold. The second worst thing is to show them you're panicked. Don't add more stress to the mix, because stress will impair learning, too!

Another obstacle to sleep that you should be aware of is the bright LED light of a computer screen, which should be turned off about an hour before bedtime to relax the overstimulated eyes and brain. In 2012 a study released by the Lighting Research Center at Rensselaer Polytechnic Institute in Troy, New York, found that just a two-hour exposure to the self-luminous backlit displays of smartphones, computers, and other LED devices suppressed melatonin by about 22 percent. Stimulating the human circadian rhythm like this before bedtime, said the researchers, could definitely affect sleep, especially in teenagers. Not all artificial light is the same, though. Some artificial light, like natural light, can drive the circadian clock. A blue light in LEDs, for instance, is healthy and can trigger circadian rhythm, something both NASA and Russian scientists are tinkering with in experiments simulating a multiyearlong mission to Mars when astronauts' circadian rhythms would be thrown into disarray.

As a parent, you have your own problems with fatigue at night, and maybe even your own work to get done for the next day. You have a low threshold, and it's easy to get angry at your kids, so you have to be mindful of that and try to regulate your emotions. As a single parent, I couldn't just throw up my hands and stomp off and tell my spouse, "You handle this." When you have no other person to turn to and you're it, you have to approach things from a different angle with your teenager. I remember not wanting to transfer my panic to my sons when I realized they were clearly unprepared or unable to figure out how to get their work done when there was a hard-stop deadline the morning ahead. You try to tell them that next time they can't wait till the last minute, or come home without the necessary books or papers to complete the assignment. But you also can't learn for your child. There's only so much you can do. You can offer to help in preparing an outline or doing some of the research, but you also don't want to create learned helplessness or dependence in your child.

As for your teenagers, suggest they do nontech activities before bed and do the same activities at the same time every night, not only to avoid melatonin suppression from the artificial light of computers, iPads, and smartphone screens but also to habituate the body to winding down at the same time every night. Making out a to-do list upon arrival home highlights the "planning ahead" part of the evening, and can ease anxiety and therefore sleeplessness. The bed itself should also be just for sleeping; avoid associations with eating or watching television—or even homework! Finally, try to avoid arguing with your teenagers before bed. They won't sleep well, and chances are neither will you. Remember the old adage: "Never let the sun set on an argument."

6
Taking Risks

After I was a guest on National Public Radio in March 2010 talking about the teenage brain, I received an avalanche of letters and e-mail. One woman wrote to me about her grandson, with whom she'd always been close. During high school he had become a heavy user of marijuana and alcohol. He was fined for speeding on one occasion, cited for reckless driving on another, and finally arrested and charged with DUI. Like many of the other e-mails and letters I have received over the course of the last few years, mostly from women and men I've never met, her message ended with a plea: "This is breaking my heart. He is a bright, handsome, beautiful young man and he is killing himself."

Parents and teachers know that adolescents are impulsive and more prone than either children or adults to taking risks. Novelty and sensation-seeking seem to motivate every act. And if it's not about risk-taking, then it's about rebellion—against parents, against teachers, against anyone in authority. Evolutionarily, this kind of behavior makes sense. Adolescence is the time of life when the young separate from the comfort and safety of their parents in order to explore the world and find independence. Experimental behavior is actually important for adolescents to engage in because it helps them establish their autonomy. The problem for teens is that their underdeveloped frontal cortex means they have trouble seeing ahead, or understanding the consequences of their independent acts, and are therefore ill equipped to weigh the relative harms of risky behavior. No matter how evolutionarily adaptive risk-taking and adventure-seeking are in the long run, in the short run there are outsize dangers.

While risky behavior has been a hallmark of adolescence for millennia, today's world poses special challenges, possibly greater than at any other point in human history. Increased access to "risk" via media, the Internet, and travel is commonplace and part of every teenager's life. In centuries past most teens were neatly tucked away on farms. Their range of movement and access to information were limited. And the environments in which they did roam were usually overseen or controlled by adults—parents, teachers, and other authority figures. That meant the potential for bad consequences of risky acts was also limited. It's important for us as parents to remember that just as there are many more bad choices available for teens today, there are many more good ones as well, and we should encourage positive information and experiences for our children.

Left to their own devices, adolescents frequently access stressful, inappropriate, even dangerous information on the Internet. It is likely that information has led to copycat self-harm behaviors such as cutting and even suicide in depressed kids. Teens are very vulnerable to the power of suggestion, and there are a lot more suggestions now at their fingertips via the computer. Similarly, statistics show that substances of abuse are much more readily available than they have been in the past, and unlike other generations, today's teens only have to send a text on a smartphone to gain instant access to a source of illicit drugs.

Scientists have a term for risk-taking, "suboptimal choice behavior," and for the most part adults chalk up the suboptimal choices of teenagers to their impulsivity, irrationality, youthful egocentricity, or pervasive sense of invulnerability. Even Aristotle weighed in on "crazy" Greek teenagers more than two thousand years ago when he wrote that young people thought and behaved differently from adults because they were "passionate, irascible, and apt to be carried away by their impulses." He also wrote that young people were slaves to their passions because "their ambition prevents their ever brooking a slight and renders them indignant at the mere idea of enduring an injury." In other words, he concluded that teenagers are so self-focused, so unreasoning, and so prone to feeling invincible that they never consider the possibility that they might hurt themselves doing something that adults would never do. Yet "irrationality," "self-absorption," and "invincibility" are labels for an adult who would do these risky things. It is hard to use these terms in the same way with a teenager.

No matter the evidence of their peculiar, sometimes infuriating behavior, teenagers are not irrational. Contrary to that popular misconception, a person's reasoning abilities are more or less fully developed by the age of fifteen. In fact, adolescents appear to be just as adept as adults in their ability to logically assess whether a certain activity is dangerous or not. This is why teens can, in fact, get very high scores on aptitude tests, such as the SAT, which relies wholly on logic and rational deduction.

So why do teens do some of the crazy things they do? In general, teen brains get more of a sense of reward than adult brains, and as we learned earlier, the release of, and response to, dopamine is enhanced in the teen brain. This is why sensation-seeking is correlated with puberty, a time when the neural systems that control arousal and reward are particularly sensitive. But because the frontal lobes are still only loosely connected to other parts of the teen brain, adolescents have a harder time exerting cognitive control over potentially dangerous situations. Adults also are better able to access a network of frontal brain areas than adolescents, whose brain regions engage in more "connectivity" to assess risks, rewards, and consequences.

In a study of 245 people, ages eight to thirty, University of Pittsburgh researchers monitored the ability of subjects to inhibit their eye movements. Instructed to look at a light on a screen in a dark room, the volunteers were told to look away from a second, flickering light when it appeared. The natural tendency of the brain is to be curious and to follow the novel information—especially if it is forbidden. This response inhibition, as it is called by psychologists, is poor in children and much better in adolescents. In fact, by the age of fifteen, if teenagers are sufficiently motivated, they can score nearly as well as adults. What fascinated the Pittsburgh scientists was the difference in brain scans between adolescents and adults. Although adolescents scored similarly to adults, adults used far fewer brain regions but could engage their frontal lobes, and this made them better able to resist temptation. Hence, adolescents had to put much more effort into staying away from what was forbidden.

In another unusual brain-scanning experiment, scientists at Dartmouth College showed that adolescents use a more limited brain region and take more time than adults—about a sixth of a second more—to respond to questions about whether certain activities, like "swimming with sharks," "setting your hair on fire," and "jumping off a roof," were "good" ideas or not. Adults in the experiment appeared to rely on nearly automatic mental images and a visceral response to answer the questions. Adolescents, on the other hand, relied more on their ability to "reason" an answer. The ability to quickly grasp the general contours of a situation and make a good judgment about costs versus benefits arises from activity in the frontal cortex, the same areas that we keep coming back to, the parts of the brain that are still under construction during adolescence.

Adults are also better at learning from their mistakes, courtesy of areas in and around the frontal lobes including their developed anterior cingulate cortex, which can act as a kind of behavioral monitor and help detect mistakes. During fMRI experiments, when adult subjects make an error, their cingulate cortex lights up as if to say, "Oh boy, I'd better make sure not to do that again." This part of the brain is still being wired in teenagers, making it more difficult for them, even when they recognize a mistake, to learn from it.

This is what I tried to explain to one woman who read an article I'd written about the teenage brain and e-mailed me in April 2011 about her eighteen-year-old daughter. "She is a great teen, but never thinks things through," the woman wrote. "Her friends, coaches, teachers, etc. all love her. She has a big heart and really wants to do all the right things but it does not always work out that way. I have been on her about everything from smoking, drinking, etc. It always seems she needs to learn from her mistakes."

The chief predictor of adolescent behavior, studies show, is not the perception of the risk, but the anticipation of the reward *despite* the risk. In other words, gratification is at the heart of an adolescent's impulsivity, and adolescents who engage in risky behavior and who have never, or rarely, experienced negative consequences are more likely to keep repeating that reckless behavior in search of further gratification. This reward-seeking impulse is located deep in the brain in two areas, the nucleus accumbens and the ventral tegmental area (VTA). These structures belong to the brain's pleasure center because they are responsible for releasing dopamine when a person contemplates or anticipates a reward (eating food, obtaining money, taking drugs, etc.). In effect, the nucleus accumbens both alerts us to the possibility of pleasure and motivates us when we are in a position to experience that pleasure. It turns out that this area is much more susceptible to the powers of addiction in the adolescent brain compared with the adult brain. Experimental studies in rats show that the dopamine neurons in these areas are more active and more responsive in the adolescent than in the adult. Without a fully myelinated frontal lobe to provide inhibition, this can drive risk-taking behavior higher.

Neurons, in general, fire at much higher rates in adolescent brains, making them "at the ready" to be co-opted to engage in addictive behavior. How does addictive behavior come about? It turns out that addiction is really a specialized form of memory. As we learned in Chapter 4, addiction is a form of synaptic plasticity, or LTP, except that the action is not in the hippocampus but in the nucleus accumbens and the ventral tegmental area—key areas for the reward circuit. Just like LTP and memory, addiction happens because a drug or another pleasurable stimulus strongly activates these synapses. As a result, these very plastic and active synapses respond by strengthening their connections, which causes more dopamine to be released in response to each stimulating experience. Hence the craving builds much faster in the adolescent brain than in the adult brain:

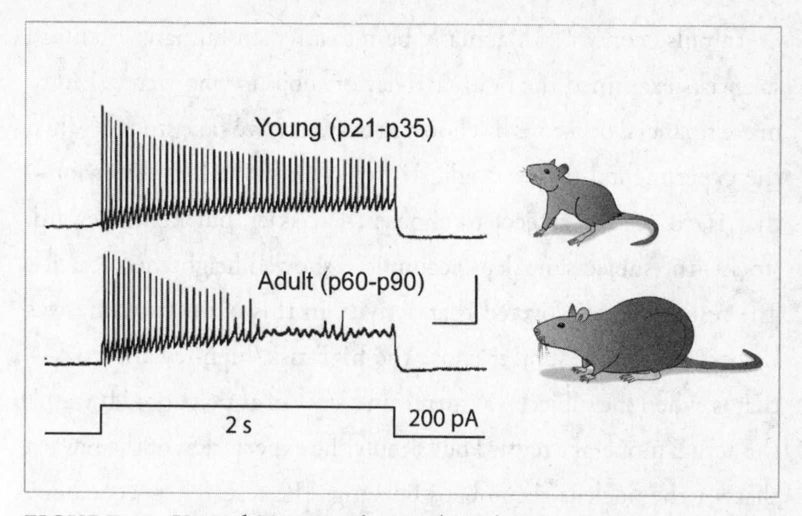

FIGURE 15. Ventral Tegmental Area (VTA) Dopamine Neurons from Young Mice Are Able to Fire More Action Potentials (pA) Than Those from Adult Mice When Stimulated: The VTA is a major part of the reward circuit, and is more active in adolescents' brains than adults'. As a result, teens have an increased propensity to seek rewards.

neurons are more active to start with and have an exaggerated plasticity in response to exposure to the addictive stimulus. Addiction, therefore, is more strongly "hardwired" into the adolescent brain, and as rehabilitation centers well know, detox is much harder and fails more often in adolescents, too. Indeed statistics show that the under-twenty-five population is the fastest-growing age group at inpatient rehab centers.

Risk and reward are inextricably linked and, not surprisingly, share many brain structures. While the nucleus accumbens and VTA house reward circuits, their activity is also controlled by the frontal lobes. In the adult, fully myelinated brain, the responsivity of these areas can be muted by the frontal lobes in the form of impulse control.

Impulse control can actually be measured in humans. Stanford scientists examined the brain activity of subjects who were asked to make financial decisions by choosing between two fake stocks. When the experimental task was adjusted to elicit risk-taking behavior—that is, to get the subject to choose the riskier, but higher-reward, stock—the subjects' nucleus accumbens showed heightened activity. The researchers discovered that activity in this brain structure was highest before the subject chose the high-risk/high-reward stock—that is, when the subject was merely in a state of anticipation. It wasn't the actual monetary reward but simply the expectation of the reward that set the nucleus accumbens buzzing. The researchers concluded that heightened positive emotions or states of arousal are indicators of a likeliness to engage in risk-taking behavior; this may be why in casinos a person who is surrounded by free alcohol and food is more likely to take a spin of the roulette wheel or a pull on a slot machine.

In another clever experiment, BJ Casey and other researchers at the Sackler Institute of the Weill Cornell Medical College showed sixty-two volunteers, ages six to twenty-nine, a series of cards depicting either happy faces or calm faces. They then asked the subjects to resist the happy faces and respond only to the calm ones by pressing the "calm" button. (The sight of a happy face stimulates the reward-seeking response in the brain in the same way the sight of a fifty-dollar bill or a tasty dessert does.) The results revealed that teens, even though they were told not to, were more likely than adults to mistakenly press the button for the happy face. Studies have consistently shown that the adolescent nucleus accumbens releases more dopamine than the adult's, so it was especially difficult for the teenage subjects to resist the "reward" of that happy face. Another factor, of course, is that they do not have the frontal lobe connectivity to send inhibiting messages to these reward centers.

Not surprisingly, these adolescent subjects reported a "greater intensity of positive feelings" during the "win" conditions. The bigger the potential payoff, the more intense the positive feelings, and the more intense the positive feelings, the greater the release of dopamine in the nucleus accumbens. Because adolescents are hypersensitive to dopamine, even small rewards, if they are immediate, trigger greater nucleus accumbens activity than larger, delayed rewards. Immediacy and emotion, in other words, are linked in the decision to take a risk and in the teen brain's inability to delay gratification.

One thing you can do as a parent, guardian, or educator to help adolescents avoid giving in to the immediacy and the emotion of rewards is to talk to them about different kinds of risky behavior. Whether it's drug experimentation or car racing, help them visualize the costs versus benefits through an analogy. Let's say you want to underscore to your teenage son or daughter that *no* possible payoff of a risky behavior is worth the chance of death; then ask something like, "Would you pick up a gun and play Russian roulette, even once, just for a million dollars?"

The dual motivation of immediacy and emotion is perhaps nowhere more engaged in adolescent behavior than around sexual activity. This was especially evident in one infamous episode about ten years ago at a New England prep school. The headline in the *Boston Globe* on Sunday, February 20, 2005, appeared on page one: "Milton Academy Rocked by Expulsions." The two-hundred-year-old alma mater of T. S. Eliot was the scene of a teenage sex scandal involving a fifteen-year-old girl who, a month earlier, had performed oral sex on five varsity hockey players, ages sixteen to eighteen, in a high school locker room. A three-day investigation was followed by the expulsions of all five boys, a leave of absence and eventual transfer of the girl, and months of media scrutiny for the prestigious old boarding school outside Boston. The school's spokeswoman, Cathleen Everett, told reporters the boys' actions were "outside common norms." At the same time, she also asserted, "Unfortunately, adolescents make big mistakes."

The Milton Academy incident was hardly the first at an elite private boarding school—it wasn't even the first at Milton Academy—but it did spawn a bestselling nonfiction book based on the case, *Restless Virgins: Love, Sex, and Survival at a New England Prep School*, written by two recent Milton Academy graduates. In the book, Abigail Jones and Marissa Miley write how teenagers today no longer regard oral sex as "an intimate act between two established partners" but rather consider it "part of a larger high school culture in which sex and girls' deference to boys reigned."

Despite adolescents' acceptance, even expectation, of sexual activity—close to two-thirds of all high school students report having sex before they graduate, according to the Centers for Disease Control—it remains a high-risk venture precisely because adolescents disregard the risks associated with sex. Although between 80 and 90 percent of teenagers report using contraception, nearly a third of girls between the ages of fifteen and nineteen who rely on oral contraceptives admit they don't take the pill every day. And among men of the same age, only about half reported they always used a condom.

It should come as no surprise, then, that risks to teens from sexually transmitted diseases are considerable. About three million adolescents every year contract one or more of these diseases. The most common STDs among adolescents and young adults between twenty and twenty-four are human papillomavirus (HPV), trichomoniasis, and chlamydia. Although teens and young adults represent just a quarter of the sexually active populace, they account for nearly half of all new STD cases. In 2004, that number topped nine million for this age group. In their book about the sex scandal at Milton Academy, Jones and Miley write about the obliviousness, or willful denial, of some adolescents of the consequences of risky sexual practices: "Teens like them—privileged, intelligent, going somewhere—didn't get HPV, herpes, chlamydia, or HIV."

The role of peers should not be underestimated when it comes to risk-taking behavior in teens. The risk-reward system in the adolescent's limbic region works closely with nearby brain structures involved in processing not only emotions but also social information. In her 2009 dissertation for a PhD in educational psychology from Temple University, Kathryn Stamoulis studied adolescent online risk-taking. The basis of her research was a survey of 934 American teens conducted by the Pew Internet and American Life Project. Stamoulis found that social isolation for girls and a lack of extracurricular activities for boys increased risk-taking behavior. In other words, socializing with friends or playing team sports appeared to have protective value in keeping teens out of risk-taking trouble. In the past, decision theorists, especially those who deal with models of economic decision-making, have often neglected the role of emotion. But how emotion factors into behavior when it comes to risk-taking is not simply a matter of degree, meaning the more emotion, the more likely it is for someone to take a risk. Mood, physiological arousal, and discrete emotional states like anger, fear, and sadness can be incidental or monumental when it comes to making a decision. What *is* key is that the brain areas involved in the perception of risk and the evaluation of rewards are closely related to the region that regulates behavior and emotion.

So here's the paradox: Adolescence is a stage of development in which teens have superb cognitive abilities and high rates of learning and memory because they are still riding on the heightened synaptic plasticity of childhood. These abilities give them a distinct advantage over adults, but because they are so primed to learn, they are also exceedingly vulnerable to learning the wrong things. How does this happen? It all goes back to the brain's craving for rewards, and the fact that anything that is learned, good or bad, that stimulates the production of dopamine is construed by the brain as a reward. This means a little bit of stimulation to a teenage brain whose synapses are firing all over the place leads to a craving for more stimulation that can, in certain situations, result in a kind of overlearning. The more commonly known name for this overlearning is addiction.

7
Tobacco

In adults, we naturally think about the physiological consequences of tobacco use, chiefly cancer and emphysema, and then simply apply those fears to our adolescents. But what we keep learning is that the adult and adolescent brains are two different things and therefore the influence of behaviors such as smoking on the teenage brain is more complex and the consequences are particularly pernicious. One of the surprising things I've learned about sleep deprivation and teenagers is that it can lead to increased cigarette use. More surprisingly still, I learned that cigarette smoking can cause a variety of cognitive and behavioral problems, including attention deficit hyperactivity disorder and memory loss, and it has been associated with lower IQ in teenagers.

While smoking has actually decreased as a favored form of substance abuse in teens in the past decade, probably owing to the now ubiquitous health warnings, we can learn a lot from the ravages of smoking on prior teen generations.

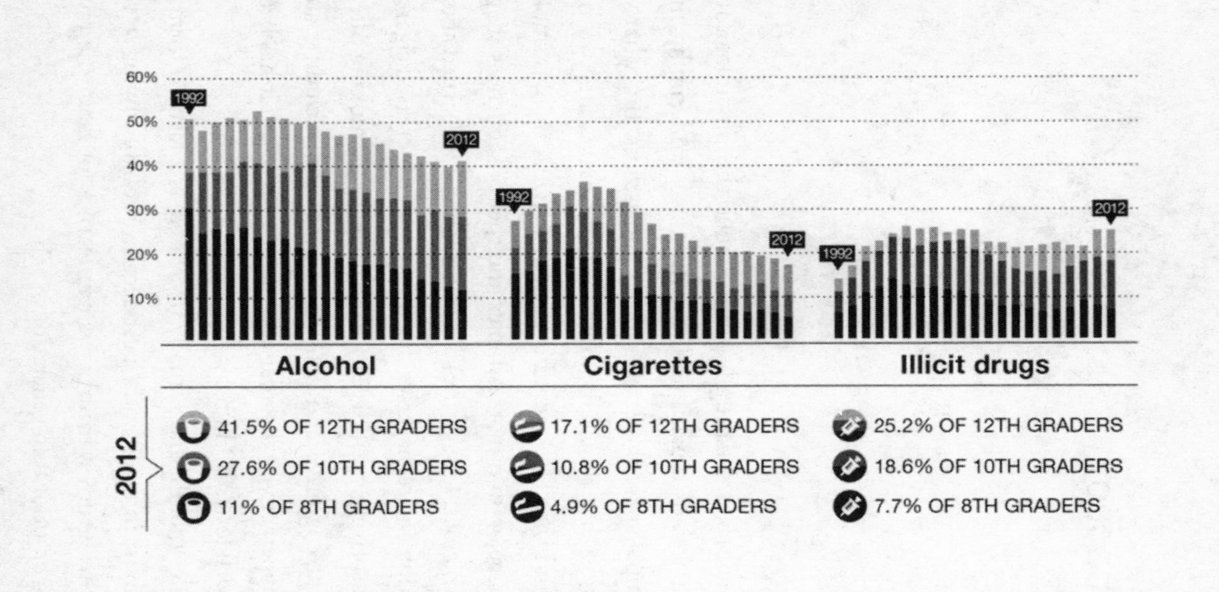

FIGURE 16. Rates of Alcohol, Cigarette, and Illicit Drug Use from the National Institutes of Health.

The fact is, teenagers get addicted to every substance faster than adults, and once addicted have much greater difficulty ridding themselves of the habit—and not just in their teen years but throughout the rest of their lives. It's as if addiction hard-wires itself into the brain when adolescents are exposed to substances of abuse. Smoking is just one example, and sadly the health toll it's taken on past generations is proof positive of the consequences. One important theme, which we will come back to in later chapters, is that because teenage brains are more plastic and primed for learning, they are, unfortunately, also more prone to addictions.

Figure 17 shows how similar the processes of learning and addiction are in the teen brain: they both arise from the adolescent brain's repeated exposure to a stimulus, which is strengthened over time. In the case of learning, the result is a good memory, and in the case of addiction, it is an increased yearning for a substance of abuse.

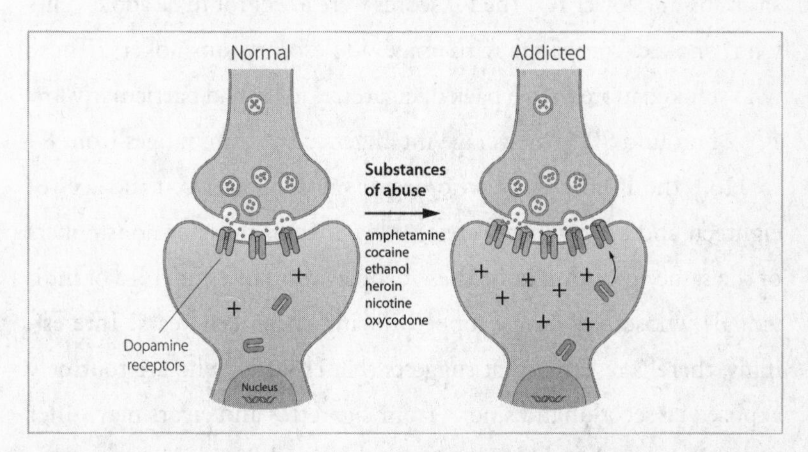

FIGURE 17. Shared Synaptic Biology of Learning and Addiction: As a stimulus, drugs have an effect on the synapses in the reward circuits of the VTA similar to that of the electric impulses in the LTP memory experiments. Both induce synaptic plasticity and reshape synapses by adding receptors; with drugs, this can lead to "craving."

A single cigarette has more than four thousand elements and chemicals in it, and many of these chemicals, depending on the amount ingested, are toxic, including arsenic, cadmium, ammonia, and carbon monoxide. Although smoking among adolescents has declined over the past fifteen years from 27 percent to 19 percent, the rate of decline has slowed in recent years. Today, 90 percent of new smokers begin before the age of eighteen, according to a new report from the US Surgeon General's office, and at least three-quarters of all adolescent smokers continue the habit into adulthood. More than 3 million high school students and more than 500,000 middle school students smoke cigarettes. It's important to remember that smoking causes lung cancer, which remains a leading cause of preventable deaths in the United States.

In an Israeli study that looked at the smoking habits of 20,000 young men in the military, researchers found a connection between smoking and lower IQ. The IQ scores were lower for male adolescents who smoked compared with male adolescent nonsmokers. Those who smoked more than a pack of cigarettes a day had particularly low IQs of around 90. (An average intelligence IQ score ranges from 84 to 116.) The IQs of those who began smoking between the ages of eighteen and twenty-one were also lower than those of nonsmokers of the same age. It might be these people are among the third of individuals whose IQs have dropped during their teen years! Interestingly, there is research that suggests that children who are routinely exposed to secondhand smoke from cigarettes and cigars may suffer not only medical problems such as asthma, colic, and middle ear disease but also damage to their nervous systems, affecting the development of their intellect and reasoning abilities. More than a third of all children in the United States are routinely exposed to secondhand smoke in their homes.

The study, by the Cincinnati Children's Environmental Health Center, used a biological marker called cotinine, a breakdown product of nicotine. In a sample of 4,399 children, ages six to sixteen, those with the highest levels of cotinine had the worst scores on tests of reading, math, and visuospatial skills, corresponding to a decrease of two to five IQ points over the control group of children with no exposure and no cotinine. Decreases in IQ were measured even at low levels of exposure to secondhand smoke.

The problem of smoking tobacco products, especially to a teenager who has never smoked before, is that it's exciting. It can offer relief from stress, which, as we will see later in this book, occurs at high levels during adolescence, and it is something communal to do with friends. To a teenager, the harmful consequences also appear to lie far down the road and therefore out of sight. Remember that recent studies have shown that the frontal lobes, which control risk-taking, are less "connected" in teens than adults. One consistent finding in human brain imaging studies is that the more teens smoke, the less activity there is in their prefrontal cortex. Poor development or damaged development of the prefrontal cortex has been found to be a cause of poor decision-making in teens.

In controlled studies, teen smokers repeatedly show difficulty in making rational decisions about their own well-being, including the decision to stop smoking. Not only do teens have less ability to utilize their frontal lobes while young, but certain experiences and substances actually interfere with normal development and consequently leave them with issues *for the rest of their lives.* Peer pressure is certainly a factor since the problem of teenage tobacco use is complicated by the fact that it's an activity that usually takes place among groups of adolescents, especially when they're socializing.

Brain plasticity during adolescence seems to make the situation of tobacco use and addiction in adolescents that much more problematic. Some studies suggest that after just a few cigarettes, the adolescent brain begins to remodel itself and create new nicotine receptors, making quitting that much harder. In fact, researchers at the University of Massachusetts Memorial Medical Center found that just a cigarette a month for an adolescent can lead to an addiction. Dr. Joseph DiFranza, a coauthor of the study, tracked nicotine addiction among a group of more than twelve hundred middle school children for four years. DiFranza found a clear pattern of progressive symptoms among sixth graders that was related to the frequency with which the students said they smoked. After two years of the study, a third of the youths who had puffed a cigarette, even if it was only once a month, said they had little control over the habit. Three or more years into the study, a quarter of all the students who tried to stop smoking experienced withdrawal symptoms, including trouble concentrating, irritability, and sleep problems. "What happens is, when you first get addicted, one cigarette a month or one cigarette a week is enough to keep your addiction satisfied," DiFranza told National Public Radio. "But as time goes by, you have to smoke cigarettes more and more frequently. So people may be addicted for more than a year before they feel the need to smoke a cigarette every day."

Animal studies corroborate this enhanced response of the adolescent brain to nicotine. When rats were given a first exposure to nicotine and their brains were examined for activation, the adolescent

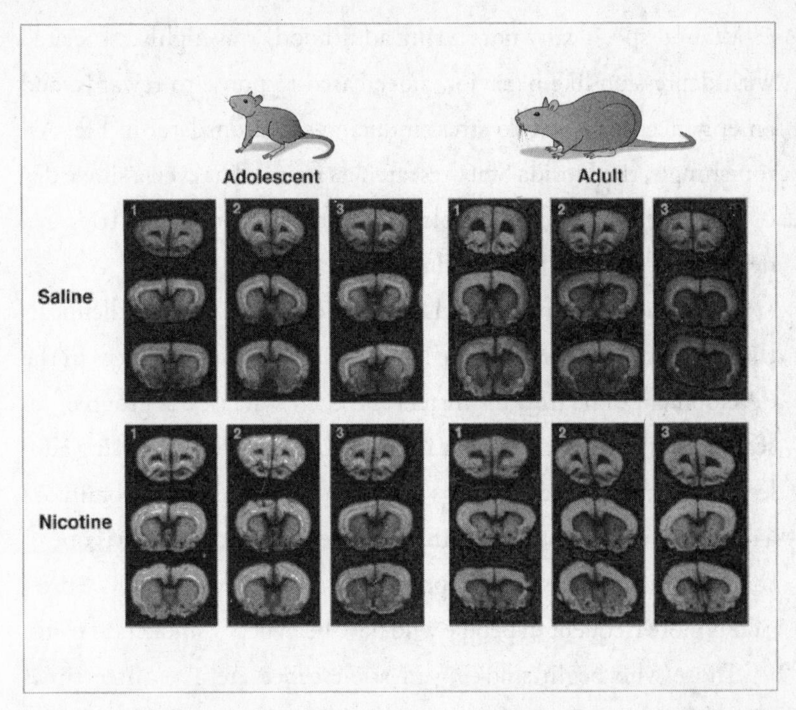

FIGURE 18. The Adolescent Brain Responds to Nicotine More Robustly Than the Adult Brain: Researchers looked at brain activation after exposure to nicotine in both adolescent and adult rats. Compared with the response to the placebo stimulus, saline, the adolescent brains "lit up" in response to nicotine, while the adults' response was minimal.

rats showed many areas lighting up (no pun intended) while the adult rats had little to no response.

Human studies have consistently shown that nicotine dependency is also high among people with mood disorders. The most recent research, in fact, indicates that cigarette smoking and dependency might actually precipitate these mood disorders, especially depression. Florida State University scientists conducted rat studies that demonstrated that a first exposure to nicotine during adolescence (and specifically not during adulthood) was highly associated with depression-like behavior, a decreased response to rewards, and an enhanced response to stress-inducing situations later in life. Astonishingly, the Florida State researchers found that even a single day of smoking cigarettes in adolescence can be enough to trigger a depression-like state later in adulthood.

Why might this be? One clue is that early exposure to chemicals that can act on the brain while it is growing can cause changes in the development of neurotransmitters and their synapses. A group of researchers at Duke showed that in rats, nicotine exposure during adolescence damaged the pathways producing serotonin in the brain. As a result, there was less serotonin, and as serotonin deficiency is one of the leading mechanisms of depression, that may explain why depression is more frequent in people who have been heavy smokers as teens.

Those who begin smoking in adolescence are also three times more likely to begin using alcohol, and long-term ingestion of nicotine has been shown to increase tolerance to alcohol, meaning it takes more alcohol to produce the same effect. Not surprisingly, smokers are ten times more likely than nonsmokers to develop alcoholism. For those teens who set off down this path, the effects of drinking are far more pronounced and the compulsion to do it again far greater than for adults who have recently decided to take this trip. Unfortunately, that compulsion, when coupled with an immature teenage cortex, frequently leads to catastrophic consequences.

So how do we steer our teenage sons and daughters away from this path? It's important, first of all, to acknowledge the attraction for adolescents. You probably did it, too, and hopefully you didn't follow through and continue after the "experiment." Teenage smoking, since the 1950s, has been an expression of rebellion against parental control and a way of bonding with a particular group of friends. It's also simply something new and different and, for that reason more than any other, tempting. Before you suspect your teenagers have started smoking, talk to them about it. Ask them, calmly, if any of their friends smoke. Affirm their own good sense not to smoke by pointing out the effects of tobacco on their growing brains. Tell them how each cigarette is hooking their brains into wanting another one and another one. Above all else, treat them with respect, acknowledging that they can learn facts. Engaging in conversation about smoking and other topics not only increases communication between you and your teenagers but also underscores their nascent sense of adult responsibility. This is also a chance to talk to them about how generations of teenagers have been manipulated by tobacco companies that portrayed smoking as glamorous in magazine ads, commercials, and movies. More pragmatically, you can help them count the weekly or monthly cost of cigarette smoking. You can even suggest ways around the peer pressure your son or daughter might feel when offered a cigarette, and appealing to teenage vanity is not out of bounds either. Remind adolescents that tobacco stains their teeth; makes their hair, clothes, and breath smell; and will probably leave them with a chronic cough and winded when trying to do sports. Remind them, too, of a relative or friend or a well-known celebrity who suffered severe health consequences directly related to smoking.

It's difficult for teenagers to look into the future because their brains are not yet wired to consider distant consequences, but that shouldn't stop you from bringing up those consequences and drilling them into your teens. They may dismiss you, they may put their hands over their ears or turn and walk away, but I promise you, it will register. Remember, they don't miss a thing at this age.

If all else fails and your teenager has already picked up the habit, this should be discussed with your physician, and this is an ideal time to rapidly "undo" the habit with a certified cessation program. One thing to steer clear of are e-cigarettes, as they can contain many times the amount of nicotine compared to a tobacco cigarette, and these can result in even more issues with addiction. More important than anything else, of course, is being a good role model for your teenager. You can't preach abstinence if *you're* still smoking.

8
Alcohol

One thing I learned as a parent of teenagers is that while you can try to set your own tone in your own household, you are really sharing parenting with all the parents of your kids' friends—adults you might not have otherwise chosen to be intimate role models for your kids. This can be a problem! For instance, an overenthusiastic divorced dad of one of my son's friends decided to win popularity points with his son by hosting a teen party for which he supplied cases and cases of beer. (Maybe his frontal lobes had not fully myelinated!) I can leave it to the readers' imagination as to what happened at the party, which wasn't a mystery, certainly, to the parents who arrived to pick up their offspring. (Many of the partygoers did not yet have their driver's licenses, and there is a midnight curfew in Massachusetts for under-eighteen drivers.) The alcohol on their kids' breath was a less-than-subtle hint as to what had happened. Fortunately there were no untoward events or mishaps at the get-together, but the incident should serve as a reminder to parents that we are all in this together and that we are a community of parents trying to bring up a community of adolescents. The problem is that you simply can't control what other parents do. I used this example with my kids to explain why, when one of my boys was going to a friend's house for a late-night party or sleepover, I would call the parents of the friend, especially if I'd never met them. While I know I completely embarrassed my sons at times, I stood resolute in my own detective work.

I've been lucky. So many others have not. For instance, that very same year there was a story tucked inside the *Milford Daily News*, a small community newspaper in an old colonial town southwest of Boston. The headline was five simple words: "Taylor Meyer Laid to Rest." The details of the death were unusual, but sadly, the cause was not. Taylor was a pretty, blond seventeen-year-old, an honors student, who on the evening of Friday, October 17, 2009, was thinking only about having fun with a few of her friends. It was homecoming night for King Philip Regional High School in nearby Wrentham, and Taylor, a senior, started partying early, drinking Bacardi rum from a bottle in the basement of a friend's house. A short time later she stopped at the home of another friend, where she downed five cans of beer before arriving at the homecoming football game at halftime wearing only a tank top in the cool night air. When the game ended, at least two dozen students, including Taylor, continued to party at a nearby abandoned airport, a local teen gathering spot that had benches and fire pits and plenty of well-sheltered space to drink alcohol without being seen by either neighbors or police.

Taylor drank five more beers at the airstrip, then decided to leave to meet her cousin. Drunk and stumbling, she started off in the wrong direction, was straightened out by friends, then wandered off again. Three days later her body was found, covered with bruises and abrasions, facedown, in a muddy area about a hundred yards from where the party had taken place. Taylor had drowned. At her autopsy, her blood alcohol level was measured at 0.13, nearly twice the legal driving limit.

"If recreational drugs were tools, alcohol would be a sledgehammer," the psychiatrist Aaron White wrote in a paper for the National Institute on Alcohol Abuse and Alcoholism in 2004. Every day in the United States 4,750 young people between the ages of twelve and twenty take their first full drink of alcohol. According to the National Institutes of Health, any alcohol use by underage youth is considered to be alcohol abuse, and in 2009 more than a quarter of young people—nearly 10.5 million teens—reported taking a drink sometime in the past thirty days. Of the 10.5 million youths who had taken a drink, nearly 7 million admitted to binge drinking, and more than 40 percent of individuals who start drinking before the age of thirteen will develop alcohol abuse problems later in life, according to a report in the *Journal of Substance Abuse*.

It comes as no surprise, then, that each year approximately five thousand people under the age of 21 die as a result of drinking. In 1965 the average age when a person first used alcohol was 17.5. Now it's 14. Alcohol saturates American culture. Even when adolescents are not taking a drink, they are being exposed to it. Researchers at Dartmouth's Geisel School of Medicine in Hanover, New Hampshire, conducted a survey in which they found that teenagers who watch PG-rated movies with scenes of people drinking alcohol are twice as likely as teens who don't watch these movies, or watch only G-rated movies, to drink alcohol and even to binge-drink. This is not an entirely American phenomenon. In France, where the minimum drinking age is eighteen, teenagers organize vodka parties via Facebook. In 2011 the French city of Lyon banned the nighttime sale of alcohol (except in bars and restaurants) to discourage teenage parties.

The reality is that when adolescents drink, whether in the United States, France, or Finland, they drink a lot, often consuming four or five or more drinks in one session. Binge drinking is defined as consuming more than four or five drinks in a single session—a span of about two hours. Studies have shown that binge drinking typically begins around the age of thirteen and peaks between ages eighteen and twenty-two. The numbers jump dramatically with high school students. More than half of all high school seniors admit to having been drunk at least once, and nearly one million high school students across the country admit to being *frequent* binge drinkers.

Novelty-seeking, poor judgment, and risk-taking behavior are partly to blame for teenage binge drinking, but there is a social component as well. Scientists have found that college students tend to pattern their drinking on the amounts they perceive their peers to be consuming: if your son's college roommate downs a six-pack every night, chances are your son will, too. What's even more alarming, though, is that researchers also found that college students consistently overestimate the amount others drink. In other words, even if your son's roommate is really drinking only three beers a night, it's likely your son *perceives* his roommate to be drinking a six-pack.

As a society we want to prevent alcohol consumption by anyone under the legal minimum age of twenty-one, and we want to prevent it 100 percent of the time—which of course is impossible. Equally ridiculous is for us as a society to say once you turn twenty-one you can drink as much as you want. While we need to have laws and regulations enforced, the problem with having a hard-and-fast line of demarcation for alcohol consumption is that the brain doesn't suddenly turn on the fully wired switch when a certain chronological age is reached. Also, making anything, especially alcohol, taboo to teenagers makes it that much more attractive. Remember, the teen brain is a novelty-seeking, risk-taking machine.

This fact, however, should not be the only reason to discourage teenage drinking. Every day we read and hear stories about teens who drink and drive and get into accidents, often with tragic results, but the dangers of even moderate use of alcohol by teenagers are far more insidious and lifelong. There are two primary misconceptions we as adults often have about the harms of underage drinking. First, we tend to think an adolescent's young body and brain are not as mature as an adult's, and therefore not as equipped to handle the immediate physiological effects of alcohol. On the other hand, because of that very youth and inexperience, we tend to believe a teenager can bounce back more quickly than an adult after a bout of drinking. Teens are resilient, at the height of their physical powers, right?

No. First of all, adolescent brains, compared with adult brains, are much better at handling the sedative aspects of drinking, including drowsiness, hangovers, and the lack of coordination. The neurotransmitter GABA, which inhibits synaptic firing, is enhanced by alcohol, and researchers have discovered that GABA receptors in several brain structures, including the cerebellum (which controls motor coordination), increase in number throughout adolescence. With fewer overall GABA receptors than adults, however, teens, especially young teenagers, experience fewer of the inhibitory effects that are enhanced by higher levels of GABA in adults. Less inhibition of activity in key brain structures such as the cerebellum means less sedation, less impairment of motor skills, and fewer coordination problems. Less inhibition means greater tolerance, and greater tolerance can result in an incentive to keep drinking. Add in peer pressure and the fact that teens spend so much time in social situations and are also more likely to drink in groups, and you have a recipe for alcohol abuse.

Tolerance for the immediate effects of drinking, however, belies the devastating long-term consequences of alcohol on the adolescent brain. More and more studies are turning up evidence of damage to cognitive, behavioral, and emotional functioning. Attention deficit, depression, memory problems, and reduction in goal-directed behavior have all been linked to alcohol abuse in teens. The damage appears to be worse for girls, perhaps because their brains develop slightly earlier than boys'. Alcohol has been shown to affect the size and efficiency of the prefrontal cortex, the site of executive functioning, as well as the hippocampus, so vital to learning and memory. In fact, researchers have shown a direct correlation between hippocampal volume and the age of onset of alcohol abuse. The earlier the use—and the longer the abuse—the smaller the hippocampus. Alcohol blocks glutamate receptors that are key for building new synapses, and this explains why people who drink heavily have major memory problems.

Alcohol can directly affect the way our synapses work, especially those used for memory. Going back to our explanation of synapses and learning, and the process of LTP (long-term potentiation—see Chapter 4), we can see how this works.

Researchers used slices of rat hippocampus and measured LTP at synapses after a burst stimulus. Normally they would expect that the burst stimulus (like a period of memorization) would result in an increase in the strength of the synaptic response to a single stimulus. What they did was bathe the slice in alcohol—or, as we scientists say, EtOH (which stands for ethyl alcohol)—and then try the same

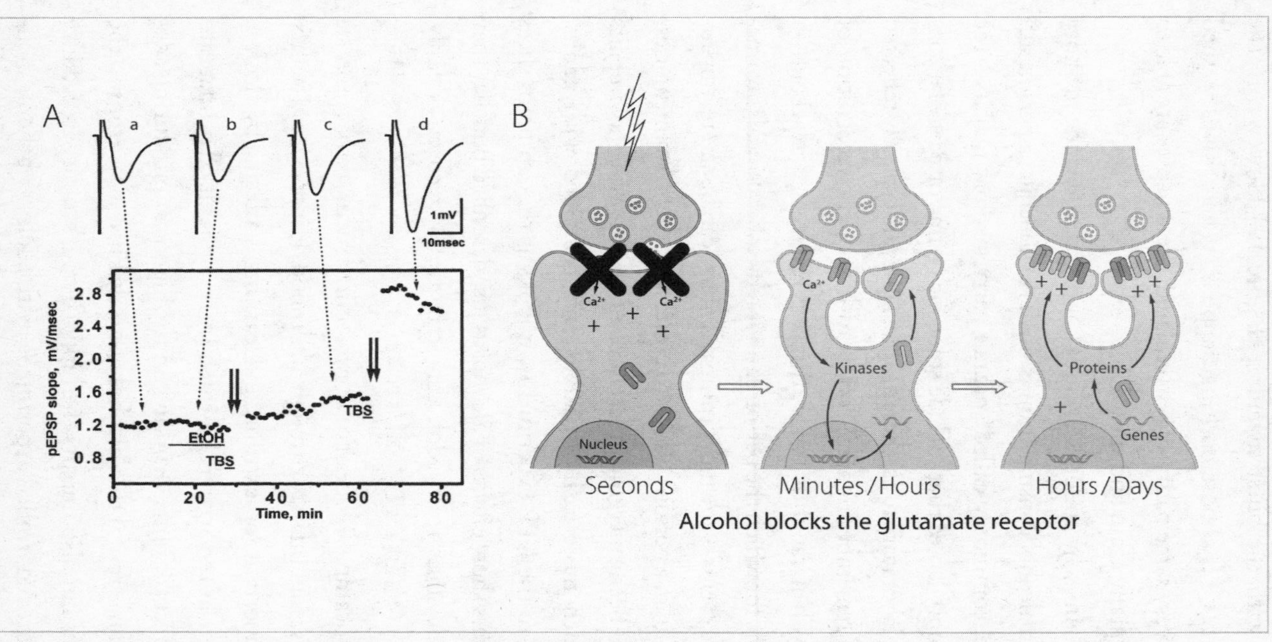

FIGURE 19. Alcohol Decreases LTP: A. Researchers looked at LTP in adult rat brains, presenting the burst stimulus both before (a) and after (b) introducing alcohol (ethanol, or EtOH). EtOH blocked LTP induction, but when it was washed out and the burst stimulus was introduced again (c), the LTP came back (d). B. Alcohol blocks glutamate receptors at the start of the synaptic plasticity process, so no LTP occurs.

experiment with the burst stimulus. (Figure 19.) However, in the presence of EtOH, almost nothing happens. When this is washed out and the very same pathway is stimulated, the synapses are able to respond normally again and show LTP.

This explains why episodes of intoxication often include memory lapses. When alcohol consumption is small to moderate, a person suffers what are commonly called cocktail party memory deficits—the kind of memory lapses that include someone's name or part of a conversation. In laboratory tests, these alcohol lapses are typically reflected in problems remembering items on word lists or the recognition of new faces. When rapid or binge drinking results in a blackout—a period of time for which the person cannot remember critical information or entire events—the hippocampal damage can be severe, impairing, in particular, a person's ability to create new long-term memories.

What has been emerging in recent research is that alcohol impairs memory much more easily in adolescents compared with adults. Going back to the LTP experiments, it looks as though brain slices from adult rats show effects of alcohol as just described but that they recover, while slices from adolescent rats cannot recover as easily. (Figure 20.)

The hippocampus is one of only two brain structures that produce new neurons from infancy through adulthood. Hippocampal neurogenesis is important for learning, and learning is affected by alcohol. Michael Taffe at the Scripps Research Institute has used primates to study the effects of binge drinking on the hippocampus during adolescence and has found that alcohol not only kills off neurons in the hippocampus but also impairs the hippocampus's ability to produce new ones. After controlled experiments with adolescent rhesus monkeys, in which four monkeys were given regular doses of alcohol equal

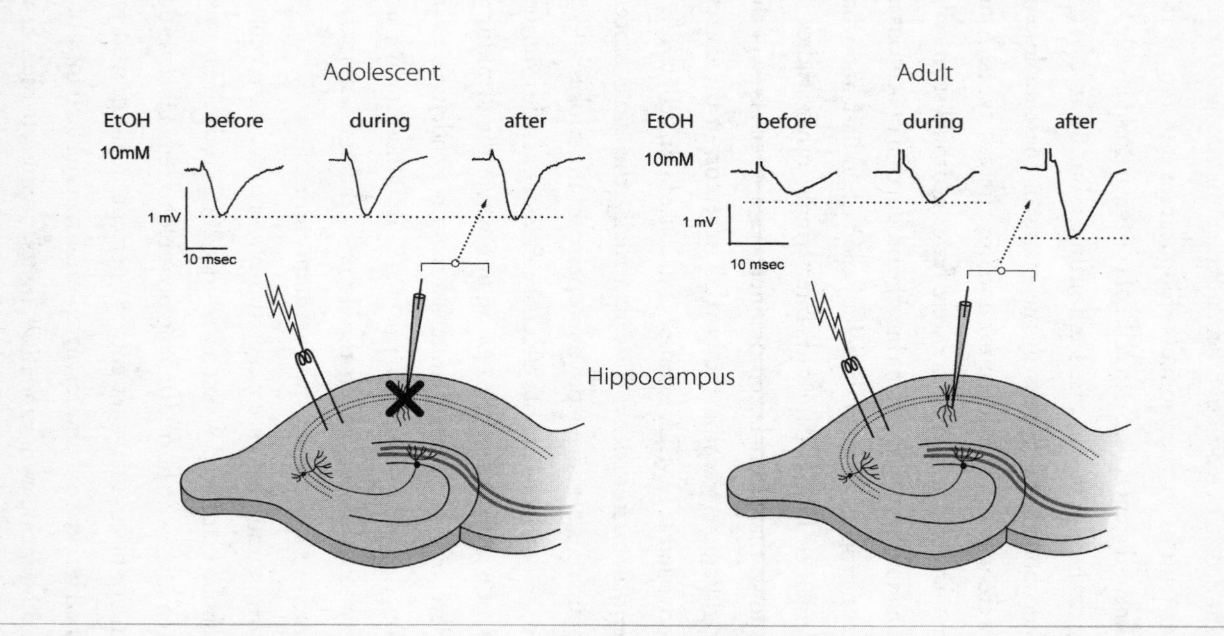

FIGURE 20. Alcohol Affects LTP in Adolescents More Than in Adults: Hippocampal slices from both adolescent and adult rats were exposed to EtOH before a stimulus burst designed to induce LTP. In both, alcohol prevented LTP, but the adult brains recovered more quickly than those of the adolescents.

to a strong cocktail for an hour a day, every day, over a period of eleven months, Taffe discovered a significant reduction not just in the number of neurons in the monkeys' hippocampi but also in the number of neural stem cells. Stem cells, of course, are responsible for generating fresh cells, in this case fresh neurons. But because of the effects of alcohol, the stem cells in the brains of these adolescent monkeys were unable to divide into more mature cell types. After just two months of drinking, there were fewer neuronal stem cells. After eleven months of heavy drinking, the production of neurons in the monkeys' hippocampi was slashed by more than half, and what neurons remained looked damaged. Interestingly, other studies in rats have shown that alcohol exposure increases a special type of the excitatory glutamate receptor, the NMDA receptor, in the cortex, and overactivation of NMDA receptors can cause brain cells to die, a process termed excitotoxicity. This is actually the same process whereby brain cells die during prolonged seizures and strokes.

Scientists talk about alcohol's effects on memory as a dose-related continuum. On one end of the continuum, where the drinking is light, the level of impairment is mild. On the other end, with severe intoxication, the deficits and impairment are profound. At any place on the continuum, alcohol affects the ability of the hippocampus to turn short-term memories into long-term memories.

An alarming number of college students say they have experienced alcoholic blackouts. When asked in a 2002 study, "Have you ever awoken after a night of drinking not able to remember things that you did or places that you went?" 51 percent of college students said they had blacked out and forgotten various events and information, including incidents of vandalism, fights, unprotected sex, driving, and spending money. An equal percentage of males and females said they experienced blackouts, but females suffered blackouts as a result of significantly less alcohol consumption. This is related to physiological differences between the sexes, with women normally having smaller overall body weight. Allowing for these differences, however, some studies suggest that given comparable doses of alcohol, women may be more vulnerable to milder forms of alcohol-induced memory impairments.

Students' drinking is a central concern of university administrations and their health centers. Many kids in college are still teenagers, and the term "in loco parentis" is used to describe the surrogate parenting that colleges attempt to provide students. There are as many ways of handling underage drinking as there are institutions of higher education in this country. Some do a better job than others, and often the ones with the strictest and most punitive policies do more harm than good. Still today at many colleges, the policy is that even if a student is not inebriated, if that student brings someone who is inebriated into his or her dorm room, the uninebriated student may still be cited by campus police for underage drinking.

Human studies of binge-drinking adolescents confirm the heavy toll alcohol takes. Rebelliousness, a tendency to engage in harmful activity, depression, and anxiety are just a few of the behavioral traits and emotional disorders that have been associated with alcohol use in adolescence. Children who begin to drink before the age of twelve also have been shown to share certain personality traits, such as hyperactivity and aggression, that may be indicators of risk for future alcohol problems. The harms caused by binge drinking don't necessarily go away when the hangover does. What scientists have discovered is that alcohol damages a specific patch of the hippocampus called hippocampal area CA1, which contains pyramidal neurons, so called because of their triangular shape. The pyramidal cells specifically help the hippocampus send autobiographical memories—memories of our experiences—into long-term storage. Alcohol blocks the ability of these hippocampal pyramidal cells to do their job, preventing the brain from forming autobiographical memories. In animal studies, researchers have consistently found that memory impairment is greater in adolescents than in adults.

Sporadic but heavy adolescent drinkers perform worse on tests of verbal and nonverbal memory than adolescents who do not drink, and adolescent girls in particular exhibit poorer visual-spatial functioning. Damaged visual-spatial functioning can cause problems in everything from doing mathematics to driving, playing sports, or remembering how to get somewhere. Adolescent boys who drink show greater deficits in attention, such as being unable to focus on something that might be slightly boring for a sustained period of time. One of the experts in this area, Dr. Susan Tapert, a psychiatrist at the University of California, San Diego, says the magnitude of difference between the adolescents who drink and those who do not is about 10 percent, which she likens to the difference between an A and a B on a school test.

Alcohol damages more than just gray matter. White matter, too, has been shown to get "dinged up" in teens who abuse alcohol. We know that white matter, the myelin sheaths that help increase the speed and efficiency of information passing through the brain, continues to develop throughout adolescence and well into early adulthood. In teens with alcohol-use disorders, the white matter of the corpus callosum, the fibers that connect the brain's two hemispheres and allow them to communicate with each other, becomes damaged, especially in an area called the splenium. Thick and rounded in shape, the splenium overlaps the midbrain, which is the part of the central nervous system associated with hearing, vision, motor control, and the sleep-wake cycle. In one study of twenty-eight teens, those who reported binge drinking had more abnormalities in their white matter than their nondrinking peers. When asked to solve a simple problem, the adolescents with alcohol-use disorders showed less activity in their prefrontal cortex than a control group and had to rely on other areas, such as the parietal cortex, to figure out the answers to the problems. One conclusion, said researchers, is that alcohol use could inhibit the ability of the adolescent brain to consider multiple sources of information when making a decision, force them to use fewer strategies when learning new information, and impair their emotional functioning. Another study showed that white matter damage increased the longer a teen drank and the more withdrawal symptoms the teen experienced.

Alcohol dependence has two common effects during withdrawal: a sluggish prefrontal cortex and a decrease in dopamine receptors, which leads to tolerance, meaning it takes increasingly larger amounts of alcohol to produce the same high. Moreover, it is likely, say researchers, that the effect of alcohol abuse on a teen's still-maturing prefrontal cortex will increase the desire for more alcohol. In fact, children and adolescents who begin drinking before the age of fifteen are four times more likely to develop alcohol dependence later in life than those who begin drinking at the legal minimum age of twenty-one.

After about a decade of solid research coming out of clinical as well as basic science studies on the effect of alcohol on teen brains, the American Academy of Pediatrics finally published a policy statement on the topic in 2010. In this statement, these experts said that schools, pediatricians, and the media needed to do a better job making the public aware of the unique vulnerabilities of this age group to alcohol. Actually making this a reality has been a challenge, and is likely to require much investment by private as well as public sources.

One of the biggest contributing risk factors for adolescents who drink is a family history of alcohol abuse. Some seven million youths under the age of eighteen are children of alcoholic parents, and researchers have found that about 50 percent of the risk of developing alcohol dependence is genetically influenced. Environment, however, counts for much of the other 50 percent. Social learning experts have found that children, especially teenagers, model their behavior on the adults who are most important to them and with whom they most frequently interact. Those who are monitored closely by their parents or guardians and who are given clear rules are less likely to abuse alcohol. In a study of three hundred teens and their parents, Caitlin Abar at Penn State found that those parents who heartily disapproved of underage drinking tended to have teenagers who engaged in less binge drinking once they got to college. Conversely, those parents who were less strict and more accepting of adolescent drinking were more likely to have teens who engaged in risky drinking behavior in college. Teens with lax parents were also more likely to surround themselves with friends who abused alcohol.

Parents' mistake, say researchers, is buying into the belief that allowing their teenagers to drink at home with friends will lead the teens to drink responsibly. Instead, says the Dutch researcher Haske van der Vorst, "the more teenagers drink at home, the more they will drink at other places, and the higher the risk for problematic alcohol use three years later."

On the flip side, there is positive news from researchers on the influence and effect of parents talking with their teens about drinking. Abar also discovered parents can shape adolescent behavior, at least while teens are still living at home. During my kids' high school years, there were so many instances of underage drinking that occurred when parents were out of the house. Once again, being a single mother, and also very aware of the legal implications, I had a lock installed on the liquor cabinet in the house. The peace of mind this gave me was priceless, especially when I left my kids alone with their visiting teenage friends or went to bed hours earlier than the sleepover gang in the basement. I only wished I could count on that being the case at the homes my kids visited. The best you can do is be that annoying parent who calls up and asks the host parents whether they will be home during the party, etc. I tried to do this behind the scenes and keep this dialogue parent to parent. I have to say, I was always grateful to have parents call me up and ask the very same question: I certainly did not take it personally, and no one should.

Alcohol and the risks and rewards of drinking should be introduced to kids slowly. They're impressionable and hungry for information of all kinds, so if we give them the necessary information about the pros and cons of drinking so they can make good decisions about alcohol, then *that* learning should take hold.

Every weekend, thousands of teenagers across the country will consume alcohol. Many of them will drink way too much; some may even pass out. All are likely to suffer some form of damage to their brains that may well be permanent. Many of those teens will also get into cars that are driven by other alcohol-impaired teens. Almost all of them will make it home again. A few, like seventeen-year-old

Taylor Meyer, who drowned in a muddy puddle at the edge of an abandoned airfield, will not. A month after the Massachusetts teen's death in 2009, police arrested a dozen intoxicated adolescents at a party not far from the airfield where they had once gathered to drink with Taylor Meyer and where the young girl had died. Many of them were wearing pink bracelets in honor of their dead friend—and drinking to her memory.

9 Pot

Much of the American public is fiercely divided about marijuana. On the one hand, many believe it to be no different from having a few beers. Marijuana is also now used medicinally as a painkiller, and it has recently been legalized in some states for recreational use as well. On the other hand, there are studies that show that smoking pot can lead to the use of harder drugs and even can stunt one's intellect. No wonder we're confused. Scientific breakthroughs showing how certain chemicals work on the brain have recently reaffirmed the risks that marijuana poses, especially for adolescents. What is not in dispute is that marijuana is a favored substance of abuse in all social strata and demographics, from the just-scratching-to-get-by inner cities to the well-heeled Hamptons. But because recreational marijuana use is legal now in at least two states, Washington and Colorado, the movement to embrace marijuana as a relatively benign mainstream drug of choice has accelerated.

While casual marijuana smoking may not bend many social mores in much of the country anymore, neuroscience is beginning to reveal that it is not as inconsequential—at any age—as previously thought. Pot is now regarded by many experts as a "gateway drug" leading to the use of more dangerous illicit substances. It impairs mental functioning and coordination and poses a threat to public safety when, for instance, individuals who are high get behind the wheel of a car.

Among the stories I've heard, one young person who began to smoke pot regularly at the age of thirteen and continued well into his twenties now says that though he has not smoked a joint for several years, he still feels as if he's in a fog. He doesn't even drive a car because he doesn't trust his attention and concentration; he struggles to keep a conversation going; and he says he suffers from anxiety, depression, and paranoia. He also says he's amazed that his middle-aged mother remembers all sorts of things about her early life, but that he can recollect only scattered bits and pieces.

More than 100 million Americans over the age of twelve admit to having tried marijuana at least once in their lives. Despite seven decades of criminalization, marijuana remains the most popular illicit drug in the world, with upwards of 200 million people using pot each year, according to the United Nations Office on Drugs and Crime. The highest use is among young people—and the age of initiation is getting younger.

In fact, marijuana is outpacing alcohol as a public health problem in teenagers. In the last five years, marijuana abuse has been responsible for almost two-thirds of admissions of teenagers, ages fifteen to nineteen, to rehabilitation centers, compared with less than a third for alcohol. The director of a rehab hospital in Connecticut that cares for a lot of clients from New York City and the suburbs told me recently that in the past five years the population at his clinic has gone from primarily adults to primarily adolescents and young people

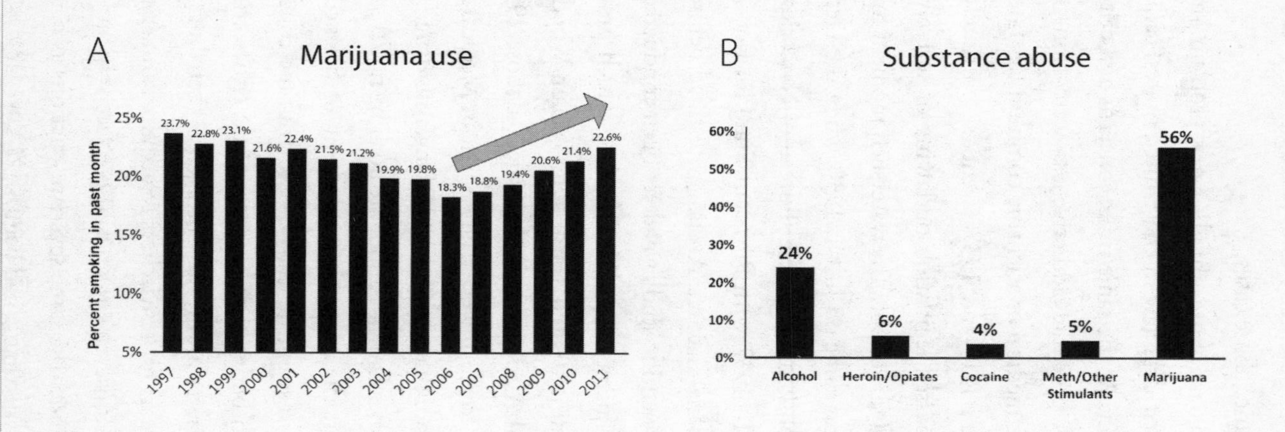

FIGURE 21. Increases in Marijuana and Substance Abuse in Teens in the Last Decade.

between the ages of seventeen and twenty-five. These people are largely multisubstance abuse cases.

Of course, marijuana abuse is hardly a new problem. In 1906 Congress passed the first drug law in the United States, called the Pure Food and Drug Act. At the same time, concerns and criticism about marijuana and its potential addictiveness were front and center, though not for the medical or scientific communities but for law enforcement. Between 1913 and 1937, twenty-seven states, beginning with California, passed legislation outlawing or severely restricting the cultivation and sale of pot. The watershed moment for marijuana in the United States came on July 1, 1930, when the Department of the Treasury established the Federal Bureau of Narcotics, headed by Harry Anslinger. In 1937, with President Franklin Roosevelt poised to sign the Marijuana Tax Act, Anslinger testified before Congress, and although no scientific studies had yet been conducted on cannabis, he said, "Marijuana is an addictive drug which produces in its users insanity, criminality, and death." Dr. William Woodward of the American Medical Association also testified at those 1937 hearings and said, contrary to Anslinger, "The American Medical Association knows of no evidence that marijuana is a dangerous drug." (At least we know that mixed messages about pot are nothing new!)

Five years later, the New York Academy of Sciences' cannabis panel issued the first scientific study on the drug and concluded there was no association between smoking pot and either criminality or insanity. Anslinger, naturally, was outraged. As retribution, he made it impossible for years for researchers to acquire cannabis in order to conduct studies. Roosevelt's tax act also effectively outlawed marijuana, placing it under the same severe restrictions as heroin and cocaine. The war on marijuana had begun.

Cannabis is the only plant that contains the unique class of molecular compounds or metabolites known as cannabinoids, including tetrahydrocannabinol (THC), which is largely responsible for marijuana's peculiar physiological and psychopharmacological effects. More than four hundred chemicals have been identified in cannabis, at least sixty of which are cannabinoids. One reason THC has such a potent effect in human brains is that we manufacture our own cannabinoids, called endocannabinoids (translated literally, "inside cannabinoids"). So we have natural cannabinoid receptors on our neurons on both sides of the synapse. When marijuana is smoked, the THC is rapidly absorbed into the blood and distributed to the tissues. Cannabis affects body temperature, blood pressure, and heart and breathing rates.

Mentally, marijuana causes an altered state of consciousness that is perceived as relaxation, pleasure, and even euphoria. Ordinary sensory experiences such as eating, listening to music, or watching a movie are intensified. Anxiety is usually decreased, but sometimes marijuana can increase it instead and can also cause depression or paranoia. At high doses users can feel a reduction in motivation and spontaneity and a general lethargy, but occasionally also confusion, hallucinations, and nausea. The immediate effects of marijuana usually kick in about fifteen minutes after smoking; others can last for three or four hours.

A consequence of pot smoking usually referred to as "the munchies" is an uptick in appetite with a neurobiological basis. Italian scientists recently isolated the probable cause: marijuana appears to affect the brain's hypothalamus, which regulates food intake.

THC is the psychoactive agent in cannabis that produces the high. The stumbling and lack of coordination experienced by someone high on pot occur because THC affects the cerebellum. Slurred speech, hyperawareness of sound, other audiological and visual distortions, and the sense of time slowing or speeding up are results of THC's effects on the sensory regions of the brain.

The pot smoked by kids today is not the pot you might have smoked in college. In 1985, THC concentrations in marijuana averaged out to less than 4 percent. In 2009 the average THC concentration in marijuana was close to 10 percent.

The most critical issue for teens is that THC disrupts the development of neural pathways. In an adolescent brain that is still laying down white matter and wiring itself together, such disruptions are far more harmful than if they were taking place in an adult brain.

The first major research breakthrough came in 1988 when Allyn Howlett and William Devane identified how TCH binds to cells, or receptor sites, in the brains of rats, and in 1990 scientists at the National Institute of Mental Health identified the location of those receptors in humans. One of the naturally occurring endocannabinoids discovered was the neurotransmitter anandamide, whose name comes from the Sanskrit word "ananda," meaning bliss. Anandamide is found throughout the brain and especially in areas involved in regulating mood, memory, appetite, pain, cognition, and emotion.

This cannabinoid, researchers found, also looks a lot like another natural compound known as the body's ideal painkiller: endorphins. Discovered in the mid-1970s, endorphins were the prize scientists had been looking for as they puzzled over why the brain has receptors for opiates. One of the most widely known opiates at the time was a chemical that oozes from poppy plants and produces sensations of euphoria and pleasure. That substance, called opium, contains naturally occurring forms of both morphine and codeine.

Parents will sometimes ask me why their teens say they need pot to relax. Part of the answer may lie in the fact that the adolescent brain, because it is firing more often and more intensely than an adult brain, is also experiencing more stress, and with increased stress comes an increased desire for relief. Enter pot. Scientists have found that THC affects the suppression of pain, and in 2011 scientists at the National University of Ireland, Galway, found that one of the keys to the suppression of pain is the production of endocannabinoids in the hippocampus. They also discovered that the hippocampus plays an active role in the suppression of pain in times of high stress. Pain suppression, or analgesia, has long been regarded by evolutionary scientists as an important biological function to help humans escape life-threatening situations.

At one time scientists believed that endorphins were also responsible for what is known as the runner's high. The general agreement today, however, is that the explanation of the sense of euphoria and relaxation that characterizes intense exercise is a lot more complicated. When researchers developed a line of mice in which the gene responsible for the production of endocannabinoids was deleted or knocked out, the mice exhibited a 40 percent decrease in running activity. If the motivation to run or to exercise intensely is related to the runner's high, then certainly endocannabinoids appear to be at least one primary driver. Evolutionarily, this makes sense, too, since someone skilled in endurance running was likely to be better at tracking down prey—or outrunning danger—than the more sedentary members of primitive societies.

Once endocannabinoids were discovered, researchers began to find them everywhere, and they are especially dense in key brain areas involved in cognition, memory, emotion, motor coordination, and motivation. When THC is introduced through inhaling or ingesting marijuana, it bathes the brain, but it also heads straight for the abundant endocannabinoid receptors in the hippocampus, amygdala, basal ganglia, cerebellum, and nucleus accumbens, where it interferes with the brain's chemical processes by either inhibiting or enhancing certain activity. Because THC binds to the endocannabinoid receptors four times as well as the brain's naturally occurring cannabinoids, THC molecules are able to overwhelm the receptor sites and interfere with normal brain function. When that happens in the cerebellum, for instance, it interrupts the smooth functioning of the motor cortex; this is why pot smokers can appear to be slack, clumsy, and slow moving and have trouble reacting promptly in dangerous situations. And that lack of responsiveness, with chronic use, results in an inability to learn from negative experiences. Most important of all, these negative effects appear to be exaggerated the earlier a person begins using marijuana. In their early teens, pot-smoking adolescents have been shown to have smaller whole brain volume, reduced gray matter, and increased damage to white matter compared with nonusers. That damage, scientists say, can still be seen in the brain scans of chronic users well into adulthood.

A small part of the amygdala, where cannabinoid receptors are especially dense, is responsible for producing the sense of awe. This same part also helps the person know what information is new and what is not so that a person can respond with the appropriate sense of heightened awareness to novel stimuli. When pot smokers claim that colors are more beautiful, music is more profound, and taste is more acute than when they're not high, it's because the flood of THC has caused this part of the amygdala to overreact. When bathed in THC, everything appears new to the brain's emotional center. Teens, whose brains are already in a nearly constant state of heightened awareness, are especially affected by this overstimulation, and breaking the pot habit, for this reason, is more difficult for adolescents than adults.

There are more problems with the amygdala the day *after* smoking pot, too. Having already been overstimulated by THC, the amygdala now has fewer cannabinoid receptor sites; this means it takes more than the normal amount of stimulation to get the brain interested in learning something new. How to do it? By smoking *more* pot, of course. For a teenager, this kind of marijuana "saturation" quickly leads to addiction.

In a blog run by the *420 Times*, a self-described "magazine of medical marijuana and natural healing," a man recently posted a message about his frustrations in trying to deal with his pot-smoking daughter:

> I'm an upper middle class, mid-thirties father. I've smoked a few times but didn't like it and don't drink alcohol either. I support legalization and usage of drugs but only if it's done in a healthy and supportive environment and by individuals that do so in a responsible manner.
>
> My 15-year-old daughter is generally an exceptional student and extremely trustworthy person. She gets straight A's in school, is very disciplined, etc.
>
> We've given her a massive amount of freedom, but we found out that she had been smoking and drinking with her friends and lying to us for the last 2 months. As of now she's effectively grounded until further notice. The issue isn't so much that she had been smoking as much as lying and hiding from us for months now.

When we told her that she can't do marijuana until further notice she broke down crying and says she "needs it" and won't stop using it. This is pretty solid evidence for me that she should not be using it.

The most significant factor in the association between marijuana abuse and potential brain damage is age. Early teen users are twice as likely to become addicted, and those who indulge in pot before the age of sixteen have more trouble with focus and attention and make twice as many mistakes on tests involving planning, flexibility, and abstract thinking. Also, the younger a pot smoker is, the more he or she smokes. Bottom line: The earlier the use, the greater the abuse.

Forgetfulness is the most widely reported cognitive deficit associated with marijuana use. THC has an effect on LTP, and can decrease the activation of glutamate receptors that build synapses during the memory process. What's more, the effects of cannabinoids can last for days in adolescents, in contrast to the briefer effect in adults. Researchers looked at LTP and simulated the effects of cannabis exposure by bathing rat brain slices in 2-AG, a synthetic cannabinoid. They found that slices bathed in 2-AG had little to no LTP, while slices in normal control conditions showed good LTP. The effect of cannabis on LTP is known to affect two points in the synapse-building process: it first prevents the signal from leaving the axon and then blocks the machinery that makes new synaptic proteins to make stronger synapses.

Memory impairment after exposure to cannabis is similar in adults and adolescents, but adolescents suffer the effects far longer. When heavy marijuana users, ages thirty to fifty-five, have been tested, studies consistently show that in the days and weeks following

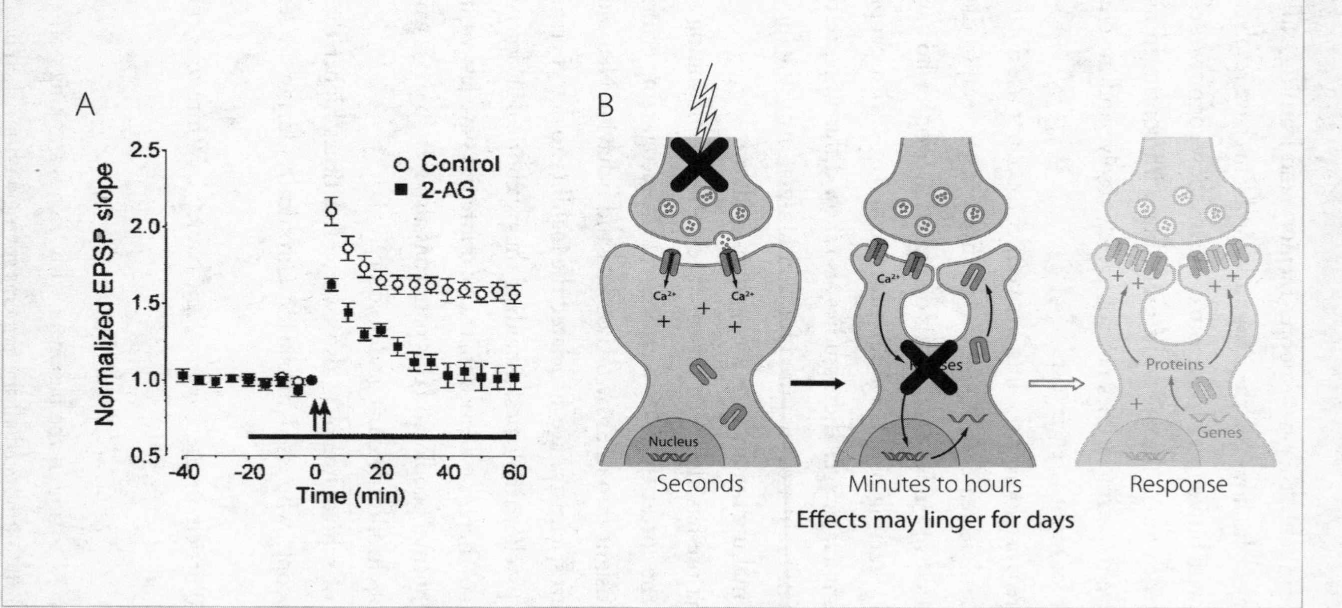

FIGURE 22. Effects of Cannabinoids on Learning: A. Brain slices stimulated in the presence of synthetic cannabinoid 2-AG show little to no LTP, while slices stimulated in normal "control" conditions showed good LTP. B. Cannabis prevents the signal from leaving the axon, and blocks new synaptic proteins from developing stronger synapses.

ingestion of cannabis, they have poorer memory and learning skills. By day twenty-eight, however, their cognitive problems have resolved. Adolescents with only short exposure to cannabis show cognitive deficits similar to those of chronic adult users, but with continued use their cognitive impairment does not completely resolve and in some cases can last for months, even years.

Even more worrisome is the link between chronic pot use during adolescence and decreases in IQ. In the last five years, several studies have shown that verbal IQ especially is decreased in people who have smoked daily starting before age seventeen, compared with people who smoked at a later age. These studies also show different patterns of activation of brain areas required for decision-making using functional MRI, making the findings even more significant.

Marijuana use has also been observed to inhibit functioning in the areas of the cortex that play a critical role in recognizing errors, specifically insight into one's own thoughts and behavior. Neurologists and neuroscientists have associated the inability to detect errors with several psychopathologies, including the psychotic delusions of schizophrenia. There is evidence that schizophrenics have less white matter than normal people, and this pattern is also seen in those who have used pot chronically during adolescence.

The risk of schizophrenia is also two to five times higher than normal in people who used marijuana chronically during adolescence.

In a 2010 article in the *Toronto Star*, Nancy J. White tells this story:

> At age 17, sitting in the basement with friends smoking pot, Don Corbeil first noticed all the cameras spying on him. Then he

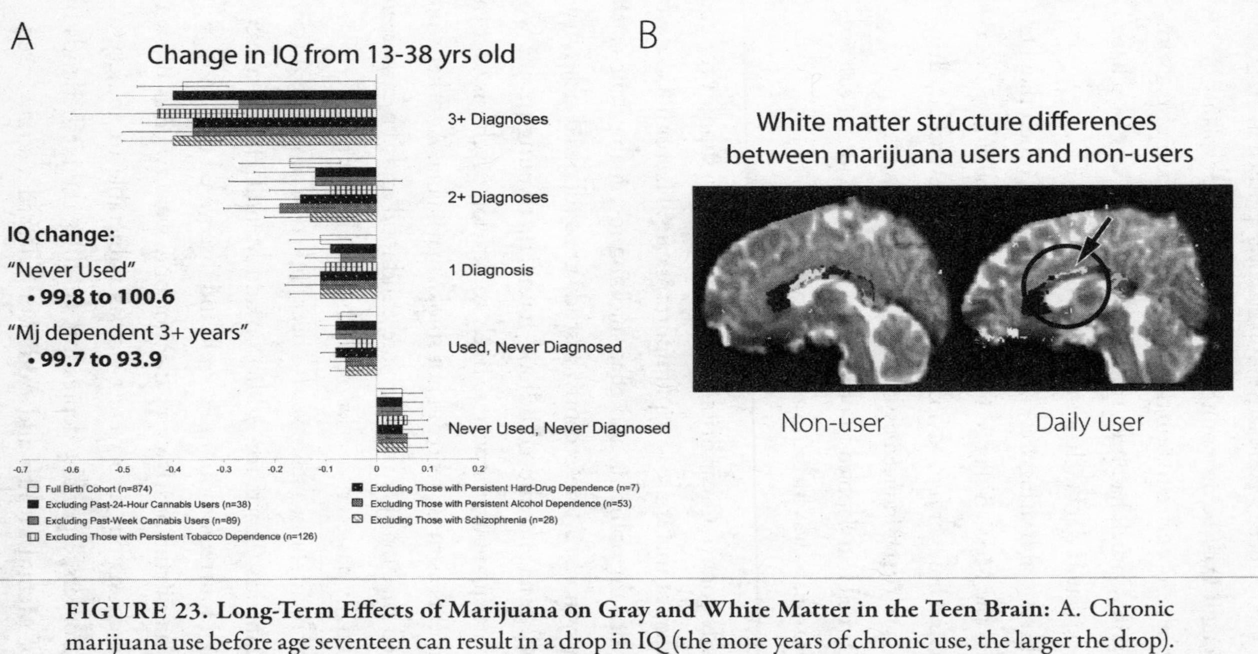

FIGURE 23. Long-Term Effects of Marijuana on Gray and White Matter in the Teen Brain: A. Chronic marijuana use before age seventeen can result in a drop in IQ (the more years of chronic use, the larger the drop). B. Imaging studies of human brains show that white matter in the corpus callosum (the area that joins the two hemispheres of the brain together) is thinner and less myelinated in those who used marijuana chronically during adolescence.

became convinced a radioactive chip had been planted in his head. "I thought I was being monitored like a lab rat," he explains.

It never occurred to him that marijuana could be messing with his brain. Corbeil had been smoking pot since he was 14, a habit that escalated to about 10 joints a day.

He started hearing voices and, at one point, Corbeil thought he was the Messiah. Police found him one day talking incoherently, and brought him to hospital, where he was eventually diagnosed with drug-induced psychosis.

Corbeil had dabbled in other drugs, such as acid and ecstasy. But marijuana was his mainstay.

A study done in Great Britain found evidence to support the conclusion that cannabis use was an important causal factor in the development of schizophrenia and that smoking marijuana doubles the risk of psychosis. More recently, a study that followed nearly two thousand teenagers into adulthood found that young people who smoked marijuana were twice as likely to develop psychosis over the next ten years as those who did not smoke marijuana. Another study concluded that marijuana use in late childhood and the early teens could actually hasten the onset of psychotic behavior by three years.

Other mental illnesses have also been linked to cannabis. A scientist at Canada's Centre for Addiction and Mental Health surveyed more than fourteen thousand people and discovered that those who smoked marijuana nearly every day not only were twice as likely to suffer psychosis but also were twice as vulnerable to anxiety and mood disorders, especially depression. In 2010, the Netherlands Institute of Mental Health and Addiction released a study based on data collected by a World Health Organization mental health study of fifty thousand adults in seventeen countries. One of the conclusions the Dutch researchers came to was that early marijuana use was linked to a 50 percent increase in the risk of having an episode of clinical depression after age seventeen. One of the largest studies followed tens of thousands of young Swedish soldiers for more than a decade. The heaviest users—that is, those who said they had used marijuana more than fifty times—were six times as likely to develop schizophrenia as those who had never smoked pot.

Why is this happening? Once again basic research is showing us that marijuana exposure during adolescent brain development can change the receptors in multiple areas of the brain—not only in the hippocampus and cortex, resulting in changes in cognition, but also in the nucleus accumbens in a way that can increase the "addictability" of the brain to other substances. This goes both ways actually. Nicotine in cigarettes also changes the number of cannabinoid receptors in the brain and makes brains more sensitive to the effects of marijuana.

Ever since marijuana entered the public discourse, people have debated whether it's a gateway drug or not—that is, can smoking pot lead to use of harder drugs? At least one expert on adolescent drug use told me he believes pot is, in fact, a gateway drug, though not in the way you'd expect. He said it's not because of peer pressure, per se, but rather because of peer exposure.

"Look, you start with pot at thirteen," he said. "When looking back at this kind of kid, when he was on pot, he was around people trying other things, and you have less of an ability to say no to other drugs because of your still-developing frontal lobes. You're probably already high anyway, so piggybacking on pot with another drug doesn't seem all that bad."

Another little-known fact is that levels of two abrasive components in marijuana smoke, tar and carbon monoxide, are three to five times greater in cannabis consumers than tobacco users. Smoking five marijuana cigarettes is equal to smoking a full pack of tobacco cigarettes, according to the American Lung Association. Marijuana smoke, which users inhale and try to hold in their lungs for as long as possible, also contains 50 to 70 percent more cancer-causing chemicals than cigarette smoke contains. (The use of a bong, a pipe that filters cannabis through water, is no insurance against these deleterious chemicals since the principal cancer-causing ingredient in marijuana is benzopyrene, which does not dissolve in water.)

All this research on the effects of marijuana on the teen brain should effectively be a public service announcement for parents everywhere, and the message is crystal clear: Adolescent brains are not as resilient as adult brains when it comes to marijuana. Teenagers are especially vulnerable to the drug because they are at a critical stage in the development of two of the most sophisticated parts of their brains—the frontal and prefrontal cortex—and these are precisely the parts most affected by marijuana. This is not minor or incidental. These brain regions are used every day for basic cognitive tasks, whether it's abstract thinking, the ability to change one's behavior in relation to changing demands in the environment, or the inhibition of inappropriate responses.

If you grew up in the 1960s or 1970s, it's more than likely you did some level of experimentation with marijuana, but you surely know by now that today's pot is not the Mary Jane of yore. It is exponentially more potent and, for that reason alone, exponentially more seductive and dangerous. So before we talk about how to steer your children away from pot and how to talk to them about its dangers, let me tell you first what you should *not* do. Do not minimize the subject—either out of your own belief that smoking a little weed didn't hurt you or because you're too afraid of learning that your teenager is indulging in marijuana. Don't even joke about past pot use with your spouse or friends, because kids do pick up on these things. They do notice. They are paying attention and they are filing it all away.

Recent research shows that fear of losing their parents' trust and respect is the greatest deterrent to adolescents' drug use. They won't tell you this, of course, but when asked by researchers what prevents them from experimenting, a majority of non-drug-using teenagers say it's because their parents expected them not to and that their parents would be disappointed in them if they did. So take advantage of this power, however unacknowledged by your sons and your daughters. Whenever possible, be concrete and practical when talking about drug use. What are their goals? What are the things they value most? Reiterate to them that college, a scholarship, making varsity, or passing drivers' ed—all those things will become more difficult to achieve if they use pot. To be convincing, of course, you also need to be knowledgeable about what marijuana does to the brain. That's where this book comes in handy. You need to know what to say when your teenagers try to argue that pot is harmless, that it makes them feel good, that it doesn't affect them negatively. For instance, if your son says that smoking pot helps him to relax and relieves him of his anxiety, then you have to remind him that he'll always feel anxiety throughout his life and he can't always turn to pot to find relief. It's important to figure out what's causing the anxiety and deal with the source rather than try to "medicate" it away.

Don't lie to your kids either. If they ask you whether you got high or smoked pot when you were their age and you did, then you must be honest. But be honest within a context. Remind them that today's marijuana tends to be more potent and easier to access given social networking, among other things. Remind them that scientists didn't know nearly as much as they do today about the effects of pot on the brain and that your kids have the advantage of this knowledge when they make choices.

And because repetition is good for the adolescent soul (even though teens complain about it), don't think that bringing up the subject or asking your son or daughter about pot smoking just once is enough, because it isn't. That's why whenever I can, I use the news or a story of a neighbor's teenager or a new scientific study as a "teachable moment," a chance to talk to my boys about what pot and cigarettes and alcohol and hard drugs have done to others and are capable of doing to them. Don't avoid talking to your kids, even if you think they aren't listening, because they are.

The implication of current marijuana research for teenagers is profound. Manipulating or interrupting this important stage of development with cannabis could change the entire trajectory of their brain development, with some deficits not appearing until much later in life. If, as parents, teachers, and guardians, we ignore the science, we do so at the peril of our own children.

10
Hard-Core Drugs

It takes only a single bad decision to result in catastrophic consequences, as it did for Irma Perez, a fourteen-year-old California girl, on April 23, 2004. According to her sister Imelda, Irma was at a party where she was offered a single pill of MDMA (3,4-methylenedioxymethamphetamine), a synthetic stimulant and mild hallucinogen also known as Ecstasy. Immediately after swallowing the drug, she became sick, "vomiting and writhing in pain," but her friends, afraid of getting into trouble, delayed calling 911 or taking her to an emergency room for hours. Instead, as Imelda wrote on www.nationalparentvigil.com, the website she helped found, they made the situation worse:

> They tried to give her marijuana, thinking it would relax her and possibly help her because they had heard it had medicinal qualities. Irma suffered for hours and when she was finally taken to the hospital the next morning, she was in terrible shape. Five days later she was taken off life support and died. . . . How did Irma actually

> die? Dr. Leslie Avery and Dr. Peter Benson, a forensic medical expert, say that Irma's brain swelled from a lack of oxygen. "Her cerebellum dissolved as her brain tried to escape its confined space," Benson said (in a *San Mateo Daily Journal* article).

Every year since 2006, hundreds of ordinary Americans have gathered for a somber, candlelit ceremony in front of the headquarters of the Drug Enforcement Agency in Arlington, Virginia. The name of the event, begun by eight couples, is the Vigil for Lost Promise: Remembering Those Who Have Died from Drugs. Those in attendance on this one night every year are mostly parents, brothers and sisters, aunts and uncles, nieces and nephews, but many are just friends who gather in remembrance of someone they loved. An overwhelming number of those whose memories are honored each year were in their teens or twenties when heroin or cocaine or prescription drugs took their lives. More than 150 names are inscribed on a virtual wall on the website of the Vigil for Lost Promise. All of them were victims of drugs; all of them are deceased.

After a decade of steady decline in drug abuse, a national survey cosponsored by the Partnership at Drugfree.org and the MetLife Foundation in 2011 revealed that the use of Ecstasy had risen 67 percent among teens. Originally patented as an appetite suppressant, Ecstasy, which is usually ingested in tablet form, is currently not approved for any medical use or treatment. It is also known as molly, which is the pure form of MDMA, and it remains popular with teenagers and young adults because it is thought to enhance sex, heighten emotions, and confer a feeling of connectedness with others. Though the government classified molly as a Schedule I controlled substance—the category for drugs with high abuse potential, no medical use, and a possible fifteen-year jail term for illegal possession—it gets used a lot at raves where electronic dance music (EDM) is played, especially in Europe. An MDMA high is called "rolling" by music ravers and is essentially a prolonged state of euphoria in motion, which goes hand in hand with electronic music and its nonstop, almost hypnotic, pulsing beat. A too-powerful form of the drug can be lethal, and because people don't know what they are getting when they buy it, they sometimes bring testing kits to raves.

The negative side effects of Ecstasy are considerable—confusion, agitation, irregular heartbeat, seizures, sleep disorders, liver and brain damage, and of course death. In fact, in September 2013 two young people died at New York City's Electric Zoo music festival and four others were hospitalized as a result of taking MDMA. Though general admission tickets for the three-day concert cost $179 a day, promoters canceled the final day because of the deaths and issued refunds. The last thing that twenty-year-old Olivia Rotondo said before collapsing in front of a paramedic was "I just took six hits of molly." Hours later she was dead.

Short of these lethal and unpredictable effects, regular use of the drug has been found to impair short-term memory and learning as well as the production of serotonin, vital to the regulation of mood. In a 2008 study, Dutch researchers found sustained damage from Ecstasy on white matter maturation and hippocampal development in both humans and animals, even in low doses. Because white matter is still being laid down in adolescence, the brains of teenagers are more susceptible than adults' brains to the destructive effects of Ecstasy. Ecstasy's toxicity to serotonin cells, in particular, suggests even more disruption of memory and mood. In rats exposed to amphetamines during their "teens" and then tested in adulthood, working memory was found to be significantly impaired and prefrontal cortex function disrupted. These effects were not seen in rats exposed to amphetamines in adulthood only. Confounding the problem for adolescents is that while they seem to have greater neuronal responsiveness to cocaine and methamphetamines than adults, they have less sensitivity to some of the physical side effects, especially motor coordination. It's no wonder that studies also show that a key predictor of addiction is reduced sensitivity to these undesirable physical side effects. An adolescent who doesn't have a negative reaction the first time he or she takes drugs is much more likely to take those drugs again—and again and again and again.

The issue with MDMA is not only the immediate effects. It turns out experimental laboratory research shows that adolescent brains exposed to MDMA undergo changes in their synapses in almost every system you could imagine. These include systems for serotonin, and a lowering of serotonin can increase the risk for depression and stress response. Stress can in turn affect glutamate receptors that modulate learning and memory.

The research into how and why drugs like Ecstasy and cocaine are so dangerous for adolescents is turning up new findings every day. The basis of many of these findings is that with a still-maturing brain, teens are especially vulnerable to drugs that work directly on the brain's chemistry. In new rat studies, adolescent brains have been found to be more susceptible to lower doses of cocaine and to suffer more severe symptoms. Their brains are also more motivated to work for the cocaine reward; this means they find it difficult to abstain, and so become addicted faster and harder than adult rats. They also relapse more easily. These results are all too pertinent for humans:

nine out of ten addicts say they first used drugs before they were eighteen years old.

Adolescents process cocaine differently from adults. First and foremost, cocaine is a stimulant, but it stimulates a greater release of dopamine in the adolescent brain than in the adult brain. Two areas in the teen brain appear to be especially sensitive to the effects of cocaine: the nucleus accumbens, which, remember, is the reward center, and the dorsolateral striatum, where habits are formed. Dopamine concentrations in these two regions of the brain are greater in adolescents on cocaine than in adults. A researcher involved with rat studies at McLean Hospital, in Belmont, Massachusetts, likened these areas to a "biochemical express lane." A similar study from Canada showed adolescent rats ran around faster than adult rats when exposed to the same amount of cocaine.

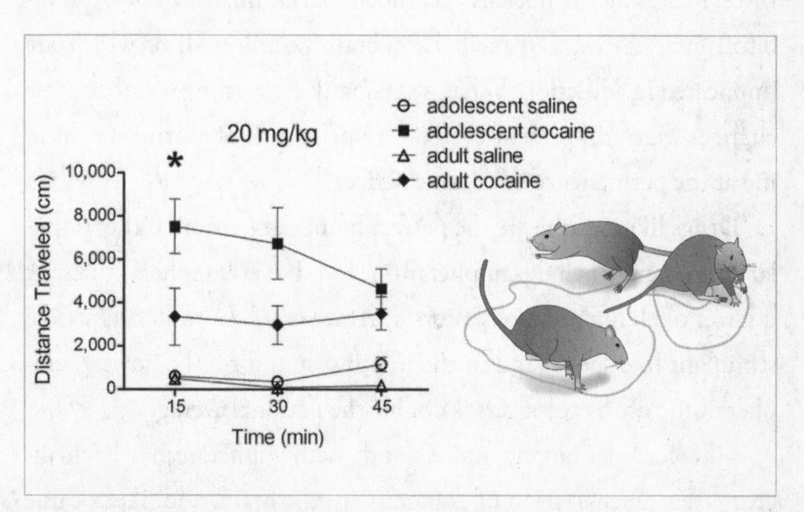

FIGURE 24. Enhanced Effects of Cocaine on the Behavior of Adolescent Rats: Cocaine makes both adolescent and adult rats run faster, but the adolescents travel farther than the adults.

Craving and relapse are hallmarks of drug addiction, and teens get hooked on drugs harder and faster than adults precisely because so many of the drugs target the very active reward system in their brains. Because the nucleus accumbens is still maturing during adolescence, its functioning is characterized by a search for the highest excitement with the least amount of effort expended. In fact, it takes less than three months for adolescents to transition from experimentation to weekly use.

Another concerning issue is that rat studies show the enhanced dopamine in the adolescent brain can actually change the way it processes information permanently, leaving the brain more susceptible to addiction in adulthood. Those researchers in Canada who showed that adolescent rats have a greater response than adults to the same amount of cocaine also found permanent changes in a host of structures, including the nucleus accumbens, striatum, insular cortex, orbitofrontal cortex, and medial forebrain bundle—all of which are implicated in addiction. What was equally concerning was that these changes were still present even after one month of abstinence, hinting at the permanence of cocaine's effect.

Drugs like cocaine are the perfect brain temptation. Other popular stimulants include amphetamines and methamphetamines. A quarter of all high school seniors in America admit to having used a stimulant like speed, and in the Southwest and rural Midwest, amphetamine use by teenagers is double the national average.

Like cocaine, amphetamines and methamphetamines both increase the concentration of dopamine in the brain, and like cocaine both produce a euphoric high. Medications used to treat attention deficit hyperactivity disorder, abbreviated ADHD, are also increasingly being abused by teenagers. Typically, teens use the pills for nonprescription purposes, most often to stay awake late at night studying or for a boost of energy or concentration when doing homework or writing a paper. But because Ritalin and Adderall and Concerta—drugs used to treat ADHD—are stimulants, they are also capable of trapping adolescents in a cycle of habitual use and addiction.

In high schools across the country, the fastest-growing drug of abuse is heroin. It didn't take long for Ian Eaccarino to become hooked on marijuana and other drugs, and he spent the better part of two years trying to get clean. On the Courage to Speak website, his mother, Ginger Katz, wrote about how quickly her twenty-year-old son became hooked on heroin.

> Nine months before he died, Ian and two friends snorted heroin for the first time. He was a college sophomore at the time. One boy became scared, one became sick—and Ian liked it. When he finally went to drug rehabilitation, he told me: "Mom, there is a smorgasbord of drugs at college. If you don't have the money, they would give it to you for free and then you're hooked."

It was Katz who found her son's lifeless body in his bed the morning after he accidentally overdosed on heroin.

> The evening before he died, I realized that he had relapsed. He knew that I was scared and that it hurt me so. He said to me, "Mom, I want to see the doctor in the morning and I don't want to move in with my friends." That was the deal. Later, he came upstairs and said, "I'm sorry Mom." It keeps ringing in my ears. Never did I think he would go downstairs and do it one more time. Even with all the remorse, the drugs were bigger than he was.

Prescription drug abuse has been on the rise nearly every year for the past decade, with 15 percent of all high school seniors reporting nonmedical use of sedatives, especially the prescription drugs Valium, Ativan, Klonopin, and Xanax. Researchers at Rockefeller University found that adolescents exposed to OxyContin (the narcotic oxycodone) can suffer lifelong damage to their brains because of permanent changes in the reward system. As the adolescent brain prunes itself, OxyContin appears to make dopamine receptors more sensitive to it. Painkillers like OxyContin activate the brain's opioid receptors and release more dopamine in the brain's reward center.

The administrator of that rehab facility for teenagers in Connecticut I mentioned earlier in this chapter said he himself became an alcoholic at the age of just thirteen. Now middle-aged, he says he has seen many young kids end up on hard-core drugs after taking painkillers for a sports injury. As a society, we're too soft on pain, he told me. Parents don't want to see their kids hurting, so they make sure their son or daughter has a prescription for painkillers. So a kid might take Percocet, for instance, and continue taking it even when there is no more pain involved because there's a bit of a physical addiction. Eventually the Percocet runs out, he explained, but you can get oxycodone on the street. Oxy, of course, is expensive, so when the person is offered much cheaper heroin, it's hard to resist. But since this person doesn't think of himself as an addict and would never shoot up, he snorts the heroin, which is ten times stronger and ten times cheaper than oxycodone. Before you know it, a kid with a sports injury is hooked on smack—and lost—and when he finally realizes there's an even cheaper high in shooting up heroin, he's hit rock bottom. It's pretty grisly, but this rehab administrator says he's seen it happen over and over again. How do you talk to teens about this kind of danger? I asked. He said they have to be ready to hear it, but the more you can get them to identify with other people it's happened to, the more you can embed the stories of addiction into situations they can relate to, and then the easier it is to get the message across.

Adolescent addiction is particularly pernicious because over a long period of usage, the brain responds to the hyperactivity of dopamine by reducing dopamine receptors, and a loss of receptors means less stimulation. The result is called tolerance. The addict must take increasingly larger doses of the drug to obtain the same high he or she experienced the first time around. And with the reward pathways so hypersensitized to being stimulated, withdrawal also comes quickly and is more pronounced than in adults, leaving the teenage drug abuser susceptible to anxiety, irritability, and depression and therefore even more determined to get high again.

The chilling realization about drug use and adolescence is this: The same brain processes that make negotiating the teen years so difficult make substance abuse more likely. An immature prefrontal cortex means less control over impulsive behavior, less understanding of the consequences, and fewer tools to stop the behavior. And an immature nucleus accumbens is also more active than an adult's, and this means teens will almost automatically seek out high-risk, high-reward activities that take little effort and offer maximum bang for the buck.

The immediate dangers of taking drugs, the life-and-death consequences, are now well known. But the ramifications of a single bad decision or impulsive act are myriad and affect not only the families of the drug abusers but also everyone else around them—including, and especially, other adolescents. After Irma Perez, the California middle school student who died after swallowing a single tablet of Ecstasy, was laid to rest, four adolescents and a twenty-year-old had to face the consequences of giving her the drug and not getting her the medical assistance she needed in a timely fashion. Two eighth-grade girls pleaded guilty to furnishing a minor with a controlled substance and cruelty to a child that was likely to result in harm or death. They were forced to cooperate with the district attorney's office during the prosecution of the three other defendants and attend an eight-month drug rehabilitation program for adolescent girls. The twenty-year-old man, Anthony Rivera, received a five-year prison sentence for providing the Ecstasy pills by selling them to seventeen-year-old Calin Fintzi, who in turn sold them to Irma's two eighth-grade friends. Rivera also provided the marijuana he thought would help ease Irma's painful reaction to the Ecstasy pill. And the fifth person, eighteen-year-old Angelique Malabey? She served six months in jail for helping Rivera hide the drugs after Irma's death.

Please understand that there is biology making substances of abuse even more irresistible to the teenager than to an adult. We need to approach the substance-abusing teenager aggressively, and possibly with more empathy than we do an addicted adult. A teen still has the capacity to change, to recover, but only with very aggressive intervention. Even a "good" kid can get mixed up with the "wrong" kids and fall into the trap of substance abuse very easily. Hence, as parents and teachers, and even teen peers, we need to keep a very watchful eye on signs of drug abuse. Withdrawal, dramatic changes in appetite and sleeping habits, excess irritability, or lack of personal hygiene, among other things, should raise concern. Talk to other adults around the teen and check if they have observed the same thing. I hate to say it, but have a low threshold for suspicion. You will be doing your child a favor. If you have to play detective in a child's room when he or she leaves for school, do it . . . for your child. If there is any evidence, you must call your pediatrician and describe what you are seeing. Addiction is a medical issue, not just "delinquent behavior"—addiction is a disease, and it can be treated. There are dozens of websites that will lead you and your teen to free care, and a good place to start if you have no other options would be the National Institute on Drug Abuse (www.drugabuse.gov). Most communities and towns have resources you can access. If you think there is something going on, you must make contact—you could be saving your child's life down the road.

11
Stress

Barely a day goes by when parents and teachers of teens don't witness an outburst of anger, tears, poutiness, withdrawal, irritability, even hostility from their mercurial kids. On the other hand, adolescents are prone to overexcitement and bursts of enthusiasm. The question often is, how do we know when things have become too extreme? How do we distinguish between episodes of normal teenage angst and something darker, more troubling? It is difficult enough to find out what our kids had for lunch at school, let alone get them to admit to depression or anxiety. And then actually getting them to articulate it, to talk about it—well, that seems like a very tall order!

Emotions are the barometer of mental well-being. It isn't hard to remember the devastation I felt when a boyfriend broke up with me in high school or I didn't get the grade I wanted in a class in college. Nor can I forget my euphoria when I learned I'd been accepted into medical school. A world without emotion is, frankly, difficult to imagine. During adolescence, more than at any other time, emotions rule our lives. Teenagers are usually up or they're down, and they are very rarely something in between. As parents we sometimes experience our teenagers' emotional highs and lows as frighteningly out of control, and because our teenagers are as yet unable to smooth things out using their frontal lobes, it's up to us to be the filter, the regulator, to provide the sense of calm their brains can't yet provide.

So when and how do we know whether an emotional outburst or mood swing, an impulsive act or even a severe disappointment, is normal teenage behavior or something we should worry about, like the first sign of depression or an anxiety disorder? There are signposts and degrees and ways of making these distinctions, but before we dive into them, you first need to understand what emotions are—and are not—when it comes to adolescent development.

In large part what makes adolescence so difficult is that much of a teenager's response to the world is driven by emotion, not reason. Adults aren't the only ones who know this; teenagers do, too. Often they describe their lives as a "drama" that can be either "too awful" or "too wonderful" depending on the circumstances. The emotional lives of all human beings are closely tied to the working of the amygdala, from which arise our most primal feelings and reactions—fear, anger, hate, panic, grief. Emotionally, the main difference between adults and adolescents is that there is much less activity in the frontal lobes of adolescents, making it harder for them to handle their emotions, especially in crisis situations.

In Chapter 1, we learned that teens may be less protected against stress compared with adults because of the way they respond to the stress hormone THP. Instead of a calming effect, the hormone stirs up additional anxiety in the teen. Stress can be induced internally, by thoughts and emotions, or externally, by the environment. Teen brains are different from adult brains in another way as well. Because teens are not fully accessing their frontal lobes, other areas of the brain can get a little out of hand and create more extreme impressions of an external threat. Primal feelings, like fear, are produced by what's called the hypothalamic-pituitary-adrenal axis (HPA). Faced with a stressful situation, the amygdala is the first to respond, and when stimulated, it releases stress hormones that signal the pituitary gland to release certain chemicals, which then prompt the adrenal gland to release adrenaline (also known as epinephrine). The adrenal glands are located right above the kidneys, and when activated in highly stressful situations, they put the body in position to respond to danger by raising the heart rate, dilating blood vessels, increasing oxygen, and redirecting blood away from the digestive tract and into the muscles and limbs in order to run away. If the instinct is to stop and defend, then our pupils dilate, our vision sharpens, and our perception of pain decreases, all in the service of preparing us for "battle." In this state of high alert, every stimulus is a possible point of danger and the body is ready for it. The threats to immediate survival are far fewer today than when our primitive ancestors roamed the earth, but the fight-or-flight response remains encoded in our genes. Adolescents, whose amygdalae are less under control by their frontal lobes, are prone to responding to situations of stress with more extreme emotions than adults, who can rely on their prefrontal cortex within the frontal lobes to control their anger and fear.

Besides adrenaline, a second neurochemical, cortisol, contributes to an adolescent's emotional cauldron. Normally, cortisol fluctuates over a twenty-four-hour cycle, with the highest levels occurring in the morning upon awakening. Those levels increase 50 to 60 percent throughout the day before dropping, at first rapidly and then more slowly, for several hours in the afternoon and evening until a low point is reached around midnight. Studies have found that in mid- to late adolescence, and especially in girls, cortisol levels are slightly higher than in the normal adult population. Negative emotions—stress, worry, anxiety, anger—have all been significantly associated with higher levels of cortisol. So, too, has loneliness; and this is why in adolescents being alone is also associated with increased anxiety and stress.

Heightened emotion goes hand in hand with stress, and the stressors for adolescents are everywhere—from speaking in front of a class to peer rejection and bullying. The effect of stressful experiences and emotional trauma on adolescents can have serious consequences for mental and emotional health later in life. Stress in adolescents works differently from stress in adults, and the effects of stress on learning and memory in teenagers can predispose them to mental health problems, including depression and post-traumatic stress disorder (PTSD). Substance abuse also often develops, when stressed kids start self-medicating by taking stimulants or antianxiety drugs they sneak out of bottles in their parents' medicine cabinets. Anxiety is astronomically high in kids today, with a host of societal issues, less consistent family life, and exposure to all sorts of stimuli on the Internet, not to mention the vagaries of social networking. These are otherwise good, normal kids, but stress can seriously strain their ability to cope.

Stress is terrible for learning. You know what I mean. A little pressure can be motivating, but once you pass beyond that, stress contributes to inattention and a real inability to learn. We have all seen a nervous kid in the midst of a spelling bee: the kid freezes, and an otherwise easy word becomes insurmountable. This "freezing" of

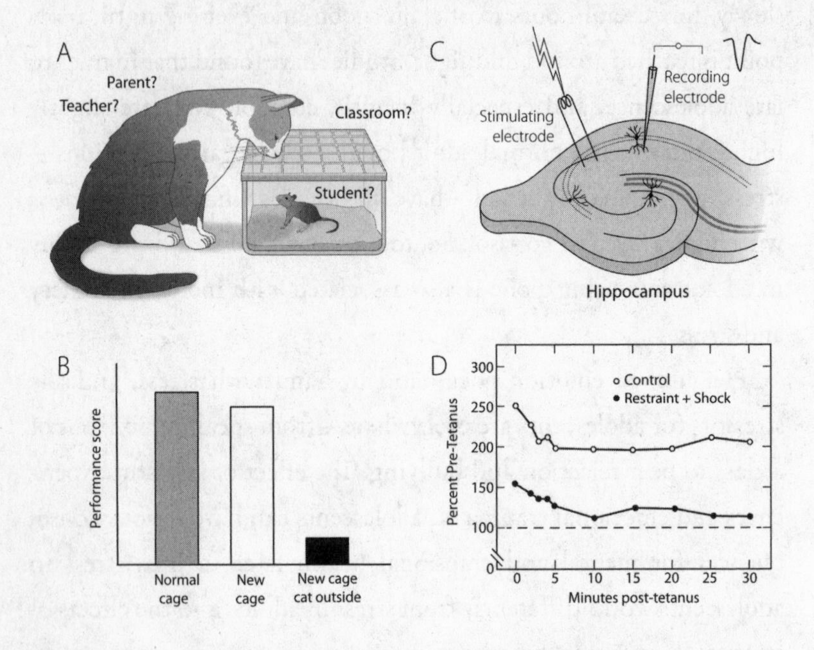

FIGURE 25. Stress Can Decrease Learning and LTP: A, B. Rats performed well when learning a task under normal home cage conditions and also when moved to a new cage. But as soon as a cat was placed outside the new cage, the rats froze frequently and were unable to learn. C, D. In hippocampal slices from the "stressed" rats, burst stimulation elicited much weaker LTP than in those of the control group, demonstrating the effect of stress hormones on synapses.

memory is real. The hippocampus, for one thing, basically stops functioning normally. Why? The surge of cortisol during a stress response can interfere with memory. Rats that are great at learning a maze under normal conditions completely freeze up and are unable to learn if a stressor, like a cat, is placed outside their cage. One of the brain structures that suffers the most damage in chronic stress is the hippocampus, critical to both memory and learning. The cascading effects on learning include impaired LTP and the elimination of synaptic connections.

We need to be mindful of what our teenagers consider stressful and realize that school can sometimes be analogous to the cage in the rat experiment, and that parents and teachers can be as stressful as the cat in that research paradigm!

Generally speaking, when this happens, thinking becomes less flexible. In animal studies, researchers have found that the brains of adults, after a period of stress, bounced back within about ten days. In adolescent animals, however, the effects of the stress were delayed by about three weeks, indicating the effects were not only long-lasting but also possibly irreversible. This gives us pause when we consider human teenagers and all the stressors that can take them over the top.

How else does stress alter the brain? Research is just beginning to yield clues. Experiments in rats show that even adolescent rats can stress out. A group of McLean researchers in Boston found that when adolescent rats were exposed to social isolation, they tended to do worse on escape tasks and showed signs of what could be construed as "helpless" behavior. Males were much more severely affected than females. When their brains were examined, they showed decreased amounts of synapses as well as myelin, especially in the frontal lobes and hippocampi. The amygdalae appeared to increase in size, perhaps in an attempt to handle the challenge. Stress was clearly altering the maturation of their brains!

Stress is a big player when it comes to emotional trauma. Adolescents are at especially high risk for experiencing emotional trauma compared with the rest of the population, and the consequences for their brain development can be devastating. One large study in North Carolina in 2010 showed that a quarter of all adolescents by the age of sixteen have experienced a "high-magnitude" event or "extreme stressor," including everything from a serious accident, an illness, or the death of a parent to sexual abuse, family violence, natural disaster, war, and terrorism. Low-magnitude events include parental separation and divorce and a breakup with a best friend, boyfriend, or girlfriend. A third of all the teens surveyed had experienced a low-magnitude event at least once in the three months prior to participating in the study.

Now that we can perform MRI scans on people and look at their gray and white matter, it seems that the same things may be happening in human teens exposed to stress as in the rats. The prefrontal cortex, hippocampus, and amygdala are major regulators of the stress response in humans, too. Like the rat's hippocampus, the teen's hippocampus appears to get smaller (not good for memory and learning), and the amygdala appears to grow in size. An increase in amygdala function may explain some of the exaggerated responses seen in PTSD.

When trauma is severe or prolonged, an adolescent is more prone than an adult to developing PTSD. Normally, PTSD develops when someone is exposed to an incident or event that threatens his or her personal safety or survival. One thing to remember is that normal adolescents, even without any abnormal stress, have exaggerated amygdala function and therefore increased stress responses. BJ Casey and her team at Cornell University's Sackler Institute used fMRI to show brain activation in response to a fearful stimulus (a picture of a frightened face) in eighty people ranging from age eight to thirty-two. Her results showed that adolescents had much higher amygdala activation when shown stressful pictures than children or adults.

So stress on top of an already overactive stress-response system can create havoc in the teen brain. Without treatment, those who suffer PTSD can become susceptible to crippling fear and anxiety

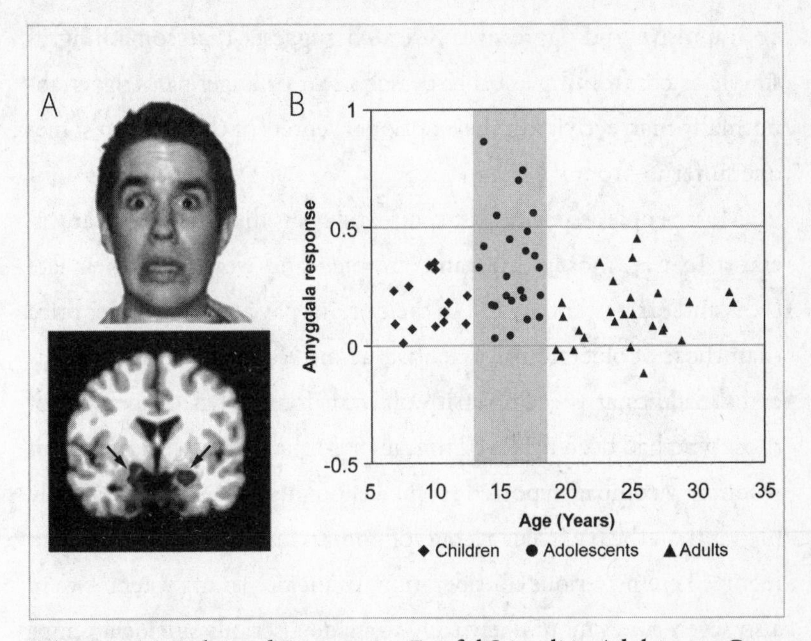

FIGURE 26. Enhanced Response to Fearful Stimuli in Adolescents Compared with Children and Adults: A. Researchers showed subjects faces with both frightened and nonfrightened expressions. B. Functional magnetic resonance imaging (fMRI) was used to determine activation of the fear circuitry in the amygdala. The adolescents' responses were generally stronger than those of either the younger or the older subjects.

throughout the remainder of their lives. The symptoms and problems associated with adolescent PTSD include not only fear and anxiety but also sadness, anger, loneliness, low self-esteem, and an inability to trust others. Behavior problems associated with adolescent PTSD also run the gamut from social isolation and poor academic performance to aggression, hypersexuality, self-harm, and abuse of drugs or alcohol.

Teenagers as well as children suffering from PTSD are likely to reenact their traumas in their artwork, with toys, or in the games they play. They are also more likely than adult sufferers of PTSD to be impulsive and aggressive. Research suggests that something as simple as confronting another person's fear or anger can trigger abnormally high activity in the emotional centers of the brain in someone suffering from PTSD.

Most people associate post-traumatic stress disorder with war veterans. In war, most combatants are men and women in their late teens and early twenties, and so their brains pay an even heavier price than those of older adults. Scientists at the National Center for Veterans' Studies at the University of Utah found that 46 percent of those who had been deployed to Iraq or Afghanistan (the majority of whom saw combat) reported suicidal thoughts. This is exponentially higher than the 6 percent average of nonveteran college students who reported giving serious consideration to suicide. Young veterans were also seven times more likely to have made a serious suicide attempt than nonveteran college students.

For adolescents in general the two strongest predictors of PTSD are exposure to violence and the sudden death of a loved one, according to the American Psychiatric Association. And of the two, the unexpected death of a loved one is the more common. Nonetheless, researchers have found that health care professionals often overlook the potential for PTSD in adolescents, perhaps because this developmental stage is already marked by emotional highs and lows and other behaviors typical of adolescence, like rebelliousness, withdrawal, acting out, and depression. All of these, however, can also be indicators of PTSD. An important difference between adolescent depression and PTSD is that fear and agitation are more characteristic of PTSD than moodiness and withdrawal. Not surprisingly, anxiety disorders are more likely to develop in children and teens exposed to trauma.

Trauma and stress are harsh on the teenage brain, but stress-induced brain alterations can also occur years before adolescence, even in utero. In one study a majority of seventeen-year-old girls and boys whose mothers had experienced above-normal stress during pregnancy (divorce, loss of a job, death of a loved one) had higher-than-normal levels of stress hormones even at rest. While males are more likely than females to be exposed to traumatic stressors, research shows that females exposed to traumatic events are more likely to develop PTSD.

Severe and chronic stress also goes hand in hand with physical and emotional abuse. Researchers at University College London used fMRI to scan the brains of twenty children and young teenagers who outwardly appeared healthy but who had been maltreated, and then compared the results with those of healthy children who had not been abused. The scientists found that during the scans, when the maltreated children and teens were shown pictures of angry faces, their amygdalae and anterior insulae, known to be involved not only in threat detection but also in the anticipation of pain, showed heightened activity similar to that in a combat soldier.

Research from late 2011 also revealed that adolescents who suffered physical or emotional abuse or neglect had evidence of brain damage, even in the absence of a diagnosable mental illness. Scientists at Yale University found that adolescents had less gray matter in the prefrontal cortex if they'd been physically abused or emotionally neglected. Reduction of activity in the prefrontal cortex in these abused youths could interfere with their motivation and impulse control, as well as their ability to focus, remember, and learn. Adolescents who were emotionally neglected also showed decreased activity in the parts of the brain that regulate emotions. Of those teenagers who were physically abused, boys revealed a greater reduction in areas of the brain associated with impulse control and substance abuse, meaning they had a greater tendency to abuse alcohol and drugs, while girls had reduced activity in the areas of the brain linked to depression. The scientists stressed, however, that these deficits were not likely to be permanent, in large part because of the brain's plasticity during adolescence.

What remains clear is that in today's world there appears to be no end of stressors in the lives of adolescents, and with their increased exposure to news events through digital technology, it is truly impossible to protect children and teens from the seemingly daily presentation of mayhem, violence, and disaster in the news. Ameliorating the effects of trauma, then, is that much more important, especially when the traumatic event is also a public one and therefore in some sense experienced by many at the same time, as happened, for instance, when two bombs exploded at the Boston Marathon finish line in April 2013. More individualized trauma is really not any different. An example would be bullying—either in person or online. Teens are very susceptible to harassment and negative criticism, and may not have the ability to see the lack of fact or logic in the accusations made by a bully. Schools and parents must take bullying seriously: it is not trivial for the victim.

The American Psychological Association suggests a number of ways to help adolescents in the immediate aftermath of a trauma that becomes public:

- Create a safe place for adolescents away from onlookers and media.
- Kindly but firmly direct adolescents away from the site of vio-

lence or destruction, the severely injured, and any continuing danger.

- Provide support to adolescents who are showing signs of panic and intense grief, such as trembling, agitation, refusing to speak, loud crying, or rage. Stay with them until they are stabilized.
- Help adolescents feel safe with supportive and compassionate verbal and nonverbal communication. Reassurances are very important.
- Provide information about the traumatic event in language adolescents can understand. This will help them to understand what happened and feel more in control.

The flip side of adolescents' vulnerability in the face of trauma is their resilience, which was also witnessed in the cases of some of the teenage victims of the bombings who attended high school proms and returned to college classrooms just weeks and months after the traumatic events. Resilience isn't something you're either born with or not. It's actually something that's learned, and for that reason teenagers, while particularly vulnerable to the negative effects of stress, are also better equipped than most adults to learn how to positively respond to stress. As an adult you are in a position to convey that information to your teenage sons and daughters, to tell them to take care, take control, and take time out. They can take care of themselves physically by eating right and getting enough sleep. They can take control of their lives by setting goals, even small ones, and working toward them one step at a time. And they can take time out from the Internet, from texting, from Facebook, and instead talk out their problems with a good listener they trust.

You have to be wise enough and mature enough to know that

this "good listener" might not be you but could be another adult figure—an aunt, an uncle, a grandmother or grandfather, even a levelheaded friend their own age. Whether or not the adult your teenager can confide in is you, what is indisputable is the importance of adults and healthy family functioning in the lives of adolescents, especially when those adolescents are experiencing extreme stress.

12

Mental Illness

Weathering teenagers' adolescence often means just riding out the rough seas with them until calmer waters are reached. But because adolescence is already a time of mood swings and behavioral irregularities, it is even more important for parents, guardians, and teachers to be aware of the emotional needs of adolescents, especially in times of crisis and stress, when adolescents' vulnerability to mental disorders is at its highest. There are two rules of thumb parents should remember: Number one, behavioral changes that seem to cluster or are associated with other symptoms should raise your level of suspicion that you might be dealing with something more than just a difficult teenager going through a phase. And number two, it is better to be safe than sorry. If you have *any* concern that radical or progressive changes are happening to your adolescent, then you must seek help for your child.

Difficult or irregular behavior in teenagers can be expressed by a variety of emotional states, from moodiness and sadness to oppositional behavior, rage, and aggression. The line between these highly charged but normal adolescent states and "real" mental illness can be difficult to determine. That's because these behavior traits (which are common in kids this age) can be seen both in teens without a diagnosable personality or mood disorder *and* in teens with one of the more severe mental illnesses, such as major depression, bipolar disorder, or schizophrenia. Signs of depression, for instance, are hard enough to detect in teenagers you're around all the time. With digital devices their constant companions, normal teens seem withdrawn compared with teens twenty years ago, making it that much harder to distinguish between a shy, introverted adolescent and a seriously depressed one. Adolescents don't engage in as many group activities as they did years ago. All this makes figuring out whether your teenage son or daughter is mentally troubled more vexing. "Real" mental illness must be diagnosed and is usually treatable, but how do you tell? When is it time to worry?

There are two general characterizations of adolescent behavior that can help in making this distinction: severity of mood and change in function. Any exaggeration or deepening of a teenager's mood swings or a predominance of one mood over another—especially anger, sadness, or irritability, and especially if it lasts longer than two weeks—is a sign of possible psychiatric problems. Changes in sleeping or eating habits, a tendency to act out more than usual, taking more risks, and spending less time with friends and family are also warning signs. So are failed friendships and an absence from extracurricular activities. Another big difference between normal but still disturbing teenage behavior and mental illness is that troubling behavior in teens without mental illness is usually isolated and, more important, doesn't interfere with the ability to function either in school or at work.

With major mood and affective disorders, however, there is rarely just one "thing" wrong. For instance, major depression is usually accompanied not only by tearfulness but also by changes in eating habits (and consequent weight gain or loss) or withdrawal from family life. Self-mutilation, alcohol or drug abuse, expressions of self-loathing, violence, and of course suicide attempts also often accompany major depression.

A sensitivity to criticism can be very acute, and given that teens already live in a world where most of what they do academically, athletically, and socially is being judged in some way, this is particularly hazardous. But only when this sensitivity is accompanied by, say, "somatic" complaints—that is, frequent pain, aches, nausea, and other physical symptoms of which they may not even be aware—is it likely they are suffering from clinical depression.

Adolescence is a unique time, as it is a period in life when some mental illnesses first emerge. What comes as a bit of a surprise is that we need to have a mature enough brain to "do" mental illness. In fact, it is known that many mood and affective disorders involve abnormal functioning of the frontal lobes, especially the prefrontal cortex. How, for instance, can you have adult-like schizophrenia that stems at least in part from abnormal frontal lobe activity if your frontal lobes aren't hooked up to the rest of your brain yet? This is likely one of the reasons why schizophrenia has its onset in the late teens and early twenties—it is not a disease of childhood.

Another interesting point is that severe mental health problems are more common in adolescents than either asthma or diabetes. One in five teens will suffer a mental or behavioral disorder serious enough to affect his or her daily life. Even more alarming: roughly half of all adult mental health disorders begin during adolescence. Among youths twelve to sixteen years of age, up to 20 percent of girls and 10 percent of boys have considered suicide. After motor vehicle accidents, suicide is the leading cause of death for teenagers and young adults. Substance abuse, high-risk behavior, plummeting school grades, even frequency of health problems can all be indicators of depression or other psychological stress in adolescents—or they can be warning signs of serious mental illness on the horizon.

Three-quarters of young adults with psychiatric illness had their first diagnosis between the ages of eleven and thirteen. In a study in England, researchers followed more than a thousand kids from childhood through age twenty-six. They found that 76.1 percent of adults who were in active treatment had a diagnosis made before age eighteen and 57.5 percent before the age of fifteen. For young adult patients who were receiving intensive mental health care, the rates were higher, with just a shade under 78 percent being diagnosed before age eighteen and a bit more than 60 percent before fifteen.

Importantly, in most cases the type of illness was the same; that is, if they had anxiety or depression as teens, then they were likely to have suffered anxiety or depression as adults. Some did not follow this pattern, though, with adult-onset schizophrenia being preceded by a wide array of juvenile psychiatric symptoms. Schizophrenia most commonly emerges in the mid- to late teens and early adulthood, including into the early thirties. However, psychosis, which can be a forerunner of schizophrenia, depression, or bipolar disorder, is seen earlier and also can be a first symptom of schizophrenia.

Also, the more minor problems of adolescent conduct disorder (CD) and oppositional defiant disorder (ODD) appeared to precede a variety of adult psychiatric disorders. While only 20 percent of eleven- to eighteen-year-olds were diagnosed with behavior problems

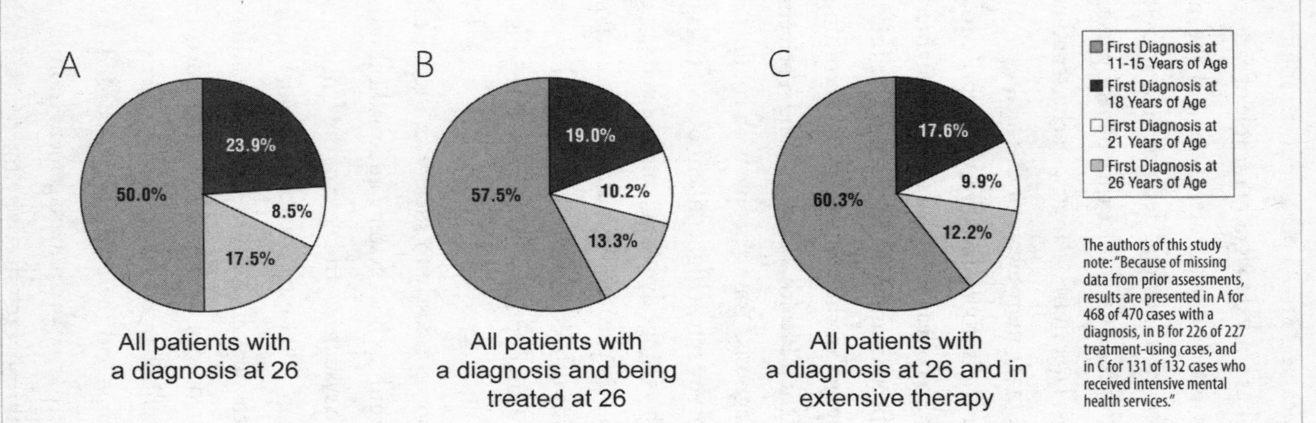

FIGURE 27. Prior Juvenile Diagnoses in Adults with Mental Disorders: A. Age at first diagnosis, of any psychiatric disorder, among patients with confirmed mental disorders at age twenty-six. B. About three-fourths of all subjects diagnosed by age twenty-six had received that diagnosis before the age of eighteen. C. Almost 80 percent of patients with very severe mental illness, requiring intensive therapy, were diagnosed by age eighteen.

in the British study, they became 25 to 45 percent of the adults with a mental illness. Psychiatry has tended to ignore the role of early conduct disorder, and this gives us reason to pay attention to it. The main point here, of course, is that even minor, and definitely major, psychiatric problems need to be addressed early since they put the person at higher risk for mental illness later in life. Parents, teachers, and even teens themselves need to be made more aware of this fact.

Conduct and oppositional disorder can be very disruptive to the family, to say the least. I have a colleague with a teenage daughter diagnosed with CD who has been in and out of residential facilities for several years. CD often emerges in the teen years, and according to the National Institute of Mental Health, between 2 and 5 percent of all teenagers receive this diagnosis. Young people with conduct and oppositional disorder also have a much higher likelihood of engaging in risk-taking behavior, such as binge drinking, unprotected sex, and driving under the influence. Fortunately, because this association is so well known, the diagnosis of a conduct or oppositional disorder should be a *huge* red flag, and an opportunity for aggressive intervention.

Conduct disorder is also very costly to society and requires expenses on the part of the family, school system, and medical system, as well as the juvenile justice system. A study out of the University of California, San Francisco, in 2008 estimated that annual medical costs for a child or teen with a conduct disorder were $14,000 compared with $2,300 for a child or teen without conduct disorder. There are now many home- and school-based programs for effective management of conduct problems, and these should be accessed ASAP once this diagnosis is suspected. In parent management training programs, therapists teach adults how to effectively interact with their children and how and when to use reinforcement or punishment to encourage appropriate behavior. Online courses are being made available for parents at places like www.thereachinstitute.org. And more and more school districts also are making available instruction and advice for both parents and teachers about how to deal with CD. CD is almost an extreme form of some of the misbehavior normally expected in adolescents.

Next up are anxiety and eating disorders, which show a surge in adolescence. As we saw in the previous chapter, the adolescent brain is more affected by stress, and hence it is no surprise that anxiety is prevalent in this age window. In fact, recent reports suggest we have an epidemic of anxiety and related disorders such as anorexia in teens these days. Studies across the United States reveal that between 2 and 9 percent of all teens have some kind of anxiety disorder. These include obsessive-compulsive disorder (OCD), panic disorder, agoraphobia, and other social phobias. There is a gender difference, with girls having higher rates and earlier onset of anxiety disorders. Anxiety disorders have a strong connection to environmental stressors; this is especially true of anorexia, the most common eating disorder, which typically appears during adolescence and is much more common in girls than in boys. I recently learned about a fifteen-year-old girl who in 2009 decided to go on a post-Christmas diet with her mother to shed a few extra pounds they'd both gained over the holidays. After five or six weeks, the mother had lost the weight she'd gained and stopped dieting, but not the daughter. In February 2010 the girl's swim coach reported his concerns to a school nurse, who contacted the parents. Apparently the girl was very good at deceiving people about how much she ate by hiding food up her sleeves or throwing it away when no one was looking. Her mother took her first to the family doctor, who agreed to monitor the girl's health, and then to a psychotherapist, whom she began to see regularly. Nothing, however, seemed to work, and in August 2010 the five-foot-seven-inch girl was admitted to the hospital weighing just ninety-one pounds. After gaining twenty-five pounds in the hospital over a period of several months, she was released to her family and sent home for Christmas. Unable to return to school until the next academic year, she took a part-time job three miles from home. Six weeks later she collapsed and was rehospitalized. That's when her parents found out that instead of taking public transportation she'd been walking three miles to work and back and not eating the lunch her mother packed for her each day. She suffered from a perforated ulcer and needed surgery, which her doctors were not sure she'd be strong enough to survive. She did, but within days her major organs started to fail. A collapsed lung, paralysis, and brain damage ensued. On March 26, 2010, she died of a heart attack at age sixteen, barely more than a year after she and her mother had decided to lose a few extra pounds.

While the acute and immediate symptoms of anorexia or binge eating and purging are troubling, the diagnoses involve another major risk: some studies show that almost half of teens with anorexia nervosa have considered suicide, and almost 10 percent have actually attempted it! Indeed, a recent study from Germany published in 2013 showed that half of all teens with anorexia had some other psychiatric diagnosis, especially depression. The researchers suggested that early treatment of the eating disorder may prevent the onset or reduce the severity of the later-life psychiatric disorder. So, just like conduct disorder, anorexia is another *red flag*. You need to get on top of it, as uncomfortable as it may feel, and alert your pediatrician about the symptoms you are seeing.

Depression is increasingly recognized as a growing problem in teenagers. The overall prevalence of depression during adolescence is higher than in childhood. Mood disorders in general, including depression and bipolar disorder as well as anxiety disorders, are the most frequently diagnosed psychiatric conditions in teens. Between 20 and 30 percent of adolescents report at least one major depressive episode, and that's enough to bump up the risk of an adult episode considerably. In fact, researchers have found that the risk of an adult episode of depression increases even if a teenager experiences only depressive symptoms and not a full-blown clinical episode.

However, adolescents and adults differ in how depression is manifested. Adolescent depression is more likely to be chronic and is associated with a thirtyfold increase in the risk of suicide. Also, while adults frequently withdraw from friendships when they are suffering from depression, teenagers often spend even more time with their peer groups. This is probably a function of the fact that teenagers not only are more social at this stage of their lives but also believe only their friends can understand the depth of their pain and suffering.

There are two other significant, postdiagnostic differences between adolescents and adults with regard to depression. More positively, depressed adolescents who take medication appear to improve more rapidly than adults and are more willing to believe in that improvement. On the downside, with regard to medication, scientists have determined that teens can react differently from adults to standard antidepressant medications, such as Prozac, Zoloft, and Wellbutrin, and have a greater risk of developing suicidal thoughts and behaviors. These drugs belong to a class called selective serotonin reuptake inhibitors (SSRIs), which boost the neurochemical serotonin in the brain.

A colleague of mine once shared the story of a friend and her husband who had two boys, both in their early teens. The older boy was depressed and put on Zoloft and then sank into suicidal despair, which they believed was a result of the medication. One day, with his younger brother in the next bedroom, he hanged himself in his closet. As he died of strangulation, his feet reflexively kicked out and banged against the wall, noises heard by his unwitting younger brother on the other side of that wall. Ten years later, at the age of twenty-four, the younger brother, who was then on antidepressant medication himself, also hanged himself. It's impossible to know if one or both boys suffered a reaction to psychiatric medicine. Even if just the older one did, the effect of having his brother kill himself in the next room could have been enough to negatively affect the younger and poison his entire future, making a suicidal depression almost inevitable.

Today the FDA includes a "black-box" label warning with all antidepressants directed at young people. Those meds include two SSRIs, Prozac and Lexapro, which were specifically approved for the treatment of depression in children and teenagers.

Sudden changes in mood, action, thought, behavior, and feelings, especially if they're severe, may signal an adverse reaction to the antidepressant and every clinician who prescribes these drugs to adolescents is required to make the risks clear to both patients and parents. Fortunately there are also atypical antidepressants that work differently from SSRIs and can be prescribed for mood disorders in children and adolescents.

Between 20 and 60 percent of adults with bipolar disorder experienced the initial symptoms of the illness before they turned twenty. Teenagers diagnosed as bipolar also present a range of symptoms different from those of adults. Adolescents have fewer episodes of pure mania and more mixed episodes of both mania and depression. Adolescents are more likely to exhibit irritability and aggressive behavior as part of the manic or depressive phase of the illness. They also have more psychotic features, such as paranoia, during the acute manic phase. Teens experience more rapid cycling than adults, with shorter durations between manic and depressive episodes, and have higher rates of dual diagnoses, suffering multiple mental health problems such as substance abuse as well as bipolar disorder.

Mania and bipolar disorder are not as common in teens as simple depression, with less than 1 percent of kids ages eleven to eighteen being diagnosed with these illnesses. However, when it is present, the onset of bipolar disorder appears most commonly in the midteens. Also, there is a gender imbalance, with depression being more common in girls than boys.

Whether antidepressant medications are involved or not, suicide remains the greatest danger of a troubled teen's life and is one of the leading cause of death in adolescence and early adulthood, according to the CDC. A nationwide survey of high school students shockingly revealed that 16 percent of kids reported at least contemplating suicide and 8 percent actually attempted it. Girls attempted suicide about three times more often than boys, but boys were more often successful, a fact thought to be due largely to their more frequent use of firearms. The rates are frightening when you think about access to the Internet, where information about how to commit suicide is sadly and distressingly available for anyone to read.

When it comes to suicide, too often the refrain from parents and teachers is that they didn't know what was going on in the teen victim's mind. That's especially difficult when an adolescent is deceptive or simply secretive about his or her innermost feelings, but that's all the more reason why we have to engage our children on a daily basis in order to know what's going on in their lives—and in their minds.

Perhaps not even that, however, could have prevented what happened to Elizabeth Shin, a nineteen-year-old college student from New Jersey. On Sunday, April 9, 2000, the day before she took her own life by setting her clothes on fire, Shin lit a few candles in her dorm room at the Massachusetts Institute of Technology in Cambridge, Massachusetts, then sat down in front of her computer and began typing in her journal. "Yoga chick," she wrote, referring to the exercises she often did to relax and take her mind off her studies. "Unfortunately I can't spend all of my life in a yoga position. Or, maybe I can?"

It was a lighthearted entry, nothing to be alarmed about, and certainly no indicator that just one night earlier she had contemplated plunging a knife into her chest. After joking about yoga, her journal suddenly turned dark. She wrote the beginning of a poem to her ex-boyfriend, who had just broken up with her, in which she asked, "May I have white roses when I die, my love? / Will you place them at the head of my grave?" Then, as if catching herself, she suddenly switched back to being the astute, objective observer, playfully making fun of herself: "Uh oh, I am in a morbid mood. I only write death poetry (bad unpoetic stuff at best) when I am morbid. . . . Here I am, typing away aimlessly, hoping to exorcise my demons. Rats. It's turning out to be more like exercising them. Are my demons in better shape than me?"

Later in the day, her parents and younger sister paid her a surprise visit, having driven up from their home in New Jersey to deliver a TV, cases of spring water, and boxes of cereal and lo mein. At dinner that night with her family at a local Chinese restaurant, she spoke about needing to get passport photos for a trip in the summer to her parents' native South Korea. She also asked her little sister to come up and spend a weekend with her soon. The family returned home. Elizabeth returned to her room, and later that night she told a friend she wanted to kill herself by taking a bottle of Tylenol with alcohol. Instead, she fell asleep. Twenty-four hours later, though, one of Elizabeth's parents picked up the phone. It was an official from MIT calling in the middle of the night. "There's been a fire," the voice at the other end of the line said.

There are few things worse than the death of a child. There are few things more horrible than the death of a child by suicide. Elizabeth Shin's self-immolation was unusual and ordinary all at the same time. It was unusual in how she killed herself; it was ordinary in that it was impulsive and yet thought out, the product of teenage angst and yet also genuine depression. Most of all, it was the result of a young woman unable to see beyond the walls of her own misery.

In 2010, I was a board member of the Society for Neuroscience, one of the largest professional groups of its kind in the world. In San Diego, at our annual meeting, I had the privilege of meeting and talking at length with the actress Glenn Close, star of such movies as *The Natural*, *Fatal Attraction*, and *The World According to Garp*. We invited Glenn to give a keynote address because of her involvement with mental health issues and advocacy of neuroscience research. Glenn is warm and affable and down-to-earth, and she has a great sense of humor. She's also tireless. In 2009 she helped found the non-profit organization Bring Change 2 Mind, which helps to foster a better understanding of mental illness. All of it, she told me and a captivated audience of neuroscientists that day, sprang from a deep well of familial commitment:

"I'm the twelfth generation of a stiff-upper-lipped, pull-up-your-socks, do-it, don't-talk-about-it, for-God's-sake-don't-show-anything, work-hard, don't-whine, make-money, don't-spend-it, win-on-the-playing-field, know-how-to-play-backgammon-bridge-and-golf, be-great-at-cocktail-parties Connecticut Yankee family," she said. "We are also a family who had absolutely no vocabulary for mental illness."

The push for Glenn came when she realized her adored sister Jessie and nephew Calen were both engaged in what she called a "life-and-death battle with bipolar disorder and schizoaffective disorder." There were signs, decades earlier, she said, when her sister suddenly developed behavioral and emotional problems in high school. Jessie, like Glenn, attended the private school Rosemary Hall (now Choate Rosemary Hall) in the early 1970s. She experienced severe mood swings and often acted impulsively when she was in her manic state, taking, for instance, a dare from classmates and sliding the dorm mother's cat down the laundry chute. The mood swings affected her studies, and she was forced to repeat the ninth grade. After she dropped out in the tenth grade, her life spiraled downward. There were suicide attempts and hospitalizations and then multiple marriages. Only when she was forty-five was she finally properly diagnosed and medicated.

Jessie also spoke at the conference and explained how when she was a teenager neither she nor her family really understood what was happening. What surprised her is that, years later, as a mother struggling with her own mental illness, she failed to recognize the symptoms in her son Calen.

"Mental illness is not easy to spot when you have no experience with it," she said. "In 1999, when Calen was sliding down into the hell of [mental illness], I thought he was simply being a trying teenager. Calen is my eldest child so I had no clue what was the norm. If only I'd known some of the warning signs. All I knew was that Calen wasn't Calen anymore. I'll always look back with shame and guilt that I had no idea what was happening to my son or what kind of help he needed."

Calen also spoke at the conference and said he thought the timing of the onset of schizophrenia is very cruel because it usually comes precisely at a moment when peer groups (that is, other teenagers) have limited empathy and self-awareness, and hence are not able to be supportive. The resulting social isolation adds insult to injury and at the worst time possible: smack in the middle of adolescence. Calen today is in his twenties, and his mother says she recognizes now, in retrospect, that her son's symptoms when he was a teenager were not adolescent angst at all but rather the first signs of an underlying mental illness.

The overriding indication that Calen's moodiness and withdrawal were not typical teenage behavior was his delusion that he was either Jesus or "the most evil thing walking the earth." Brought by his father to the emergency room of a local hospital in Helena, Montana, where they lived, he kept repeating "blue square, red square, blue square, red square . . ." as he stared at the geometric patterns of the emergency room wallpaper. It was a code, he said, to help get him back into "reality." Once he was admitted to the locked psychiatric ward, he thought he would have to fight his way out:

> I began to prostrate myself, praying for God to let me endure the fight that I would now have no choice but to take part in. For whatever reason on that day I thought I was now going to be forced

> to fight for my life. Security guards were called on to the unit as I was obviously needing to calm down. Sensing I was in danger, I grabbed the chair from the common room and stood with my back facing the wall. A nurse rushed past me to close and lock an open door, I was now cornered, and it took four guards to pin me down and restrain me by my arms and legs.
>
> When I was fighting the guards, I looked up and saw an older man standing close by with a white beard and hair. I kept begging him for help, thinking that he was God, and I didn't understand why he wasn't intervening. But it wasn't his battle, it was mine. They carried me and four points [restraints] strapped me down to a bed, finishing by injecting me with a strong dose of Haldol, and I passed out.

Schizophrenia is less common than either depression, bipolar disorder, or anxiety disorder, but it is not rare; it affects about one in one hundred people. Interestingly, the brain has to reach a certain level of maturity before it can manifest schizophrenia. The disorder commonly first shows up in a person's late teens or early adulthood. The warning signs can be similar to depression in that kids can appear withdrawn, socially isolated, and sad, with changes in their eating habits and hygiene. However, there are a few distinguishing features, which can include the presence of hallucinations, strange speech patterns, and psychosis, and psychosis, in turn, can be a highly agitated state along with paranoid behavior and delusions of persecution or grandeur. There is a schism, or detachment from reality, and that is where the word "schizophrenia" comes from. There is no actual split personality, however; schizophrenia is more a disconnection from the real world.

Schizophrenia is a chronic condition, and treatment is essential, especially when a young person is still early in the disease process.

Hallucinations should make you worry about schizophrenia, but much more common causes of hallucinations, especially in teenagers, are the drugs LSD and PCP, and even large quantities of more commonly used substances such as alcohol and marijuana. The difference is that the person who is under the influence of drugs will also show signs of sedation, lack of coordination, and confusion. Hallucinations in schizophrenia are not accompanied by these symptoms or side effects.

While stress is one of the main risk factors for schizophrenia, as it is for mood and anxiety disorders, there are at least two others: advanced paternal age at time of conception and frequent marijuana use in adolescence. Researchers at the Netherlands Institute of Mental Health and Addiction carried out a study following two thousand subjects through adolescence and found that use of cannabis in the early teens can hasten the onset of psychosis and increase the risk of schizophrenia. Those most at risk were adolescents who had an immediate family member with schizophrenia or some other psychotic disorder. Even without smoking pot, teenagers with a family history have roughly a 1-in-10 chance of developing the condition. Marijuana use, though, doubles that risk to 1-in-5. Teens with no family history, the researchers found, have a 7-in-1,000 chance of developing a psychotic illness, which doubles if they smoke pot on a regular basis.

Besides psychosis, strong negative emotions and behaviors such as extreme loneliness and apathy may be indicators of more than a passing teenage mood, especially if those emotions and behaviors last more than two consecutive weeks. If they also cause your teenager's grades to fall off the map or if he refuses to get out of bed and misses school, these, too, are indications of a possible underlying psychological disorder.

So what actually happens in the brain of an adolescent when mental illness is triggered? The chief culprit is stress, which we talked about in the preceding chapter. As adolescents' brains are maturing, their HPA axis, the hypothalamic-pituitary-adrenal axis, the body's chief stress-response mechanism, gets a workout. Researchers have found that clinical depression seems to emerge from a gradual dysregulation of the HPA axis from childhood into adolescence caused by a greater-than-normal release of cortisol in the brain. These higher-than-normal levels of cortisol both precede and predict the development of depression in adolescence and early adulthood. Why some people release more cortisol than others is not yet fully understood, and while there is no physiological or biological test for depression, researchers are hoping to develop one. Salivary cortisol can provide a good index of stress and can be collected noninvasively by having a patient drool through a straw into a collection cup while undergoing a stimulus. While not likely to be definitive, the test could become an aid in determining what psychological processes are at work in a patient who develops major depression.

Anxiety, like depression, is a frequent complaint of adolescence, and it often doesn't take much for simple nervousness, restlessness, or fear to become a full-blown anxiety disorder. Many times adolescents will describe chronic feelings of uneasiness without knowing the cause and with no apparent immediate threat or stressor. Teens, by nature, worry a lot and are prone to being restless and irritable. To be characterized as a disorder, however, the anxiety must interfere with normal functioning. Excessive worry can cause a teenager to withdraw from daily activities and become shy and hesitant to engage in new experiences. In addition, excessive worry also can push a teenager in the opposite direction, toward more risk-taking, drug experimentation, and unprotected sex as a way to overcome, diminish, or simply deny the fears. In some cases, the excessive anxiety will produce physical symptoms, too: headaches, stomachaches, fatigue, trembling or sweating, even hyperventilation.

Among adolescents diagnosed with anxiety and impulse control disorders, between 50 and 75 percent show the first signs of the disorder during adolescence. There are a number of anxiety disorder subcategories, too, as set forth by the National Institute of Mental Health, which are not specific to adolescents.

The chief difference between teens and adults who suffer from an anxiety disorder is the source of the anxiety. For adults, anxiety usually emanates from problems with health or money, difficulties at work, and family issues. For adolescents, it often has to do with friends and school—social acceptance, academic performance, etc. The difference between those with ordinary teenage angst and those with bona fide anxiety disorders is one not of content but of degree. In a 2000 study of youths being seen at an anxiety disorders clinic, patients' answers to a question about what worried them most frequently and most intensely were not much different from the answers of youths without an anxiety disorder.

Their top-five most frequent worries:

1. Friends
2. Classmates

3. School
4. Health
5. Performance

Their top-five most intense worries:

1. War
2. Personal harm
3. Disasters
4. School
5. Family

What *is* different between teenagers with an anxiety disorder and those with just normal stress is the level and constancy of the stress. In imaging studies of the brains of adolescents diagnosed with anxiety disorders, there is always more activity in the limbic system, in the fear and emotion parts of the brain, particularly in the amygdala, than in normal control subjects. Researchers have consistently found a positive correlation between amygdala activity and anxiety, but whereas depression is associated with the left amygdala, anxiety disorders are specific to the right amygdala, which is responsible for detecting emotional stimuli.

By nature, adolescents already have fairly overactive amygdalae, which means they really need their prefrontal cortices to exert even greater control. For teens at risk of an anxiety disorder, however, their still-maturing brains are not yet able to exert that kind of top-down control. For that to occur, brain regions need to "talk" to one another, and there is evidence in animal studies that adolescent brains aren't doing as much "talking" as adult brains. This is due to the relative lack of myelin covering the connecting tracks, resulting in signals not traveling fast enough between brain areas.

Adolescent girls are more liable to suffer anxiety and mood disorders than adolescent boys. Teen girls report not only more stressors in their lives than boys, but also greater distress in response to those stressors. This sensitivity may be enhanced because the typical girl, with her enlarged midline connectivity on the undersurface of her frontal lobe, is more attuned to social and interpersonal relationships at this stage. In 2009 researchers at the National Institute of Mental Health reported that just when sizing one another up, girls show more brain activity in certain emotional circuits. At a time when adolescent girls are more and more concerned with how they are viewed by their peers, interpersonal stress can play a large role in the development of an anxiety disorder.

Teenagers with high levels of anxiety, whether or not they've been diagnosed with a disorder, will often try to self-medicate. In 2011 a group of Finnish researchers were engaged in an ongoing study called the Adolescent Mental Health Cohort that included 903 boys and 1,167 girls ages fifteen and sixteen. As part of their research, the scientists looked at rates of drinking, especially in relation to those who had been diagnosed with an anxiety disorder. Four percent of the more than two thousand male and female Finns in the study had such a diagnosis. At the start of the study only 10 percent of the teens reported drinking alcohol on a weekly basis. However, after just two years, 65 percent of the anxious youths said they were weekly drinkers; this amounted to a threefold increase in risk. (The study did not look at the differences between male and female adolescents.)

The vulnerability of a teen to emotional and psychiatric issues cannot be overemphasized. The teenage years are a developmental stage whose by-products are a hypersensitivity to stress, an inability to exercise self-analysis or insight, and membership in a peer group equally unable to interpret warning signs or to offer adequate empathy. Here is a major opportunity for the adults around teenagers: Be vigilant, exercise your own well-developed skill sets to ask questions, probe, stay connected, and, most important, have a low threshold to seek medical advice or counseling for symptoms that appear to change from the ordinary. Also, as kids these days spend so much time online, isolated, warning signs could be harder to detect. Kids sit alone in their rooms on the Internet and the phone. Years ago, social isolation was quite easy to notice: a kid sitting alone in the cafeteria, on the school bus, or on the top bench of the bleachers. Now moderate physical isolation is a natural consequence of online social interaction, making it much more difficult to detect. But, as a parent, you don't know until you check. Be part of your kids' lives at home. Don't count on your teenager—or his or her friends—to sound the alarm.

13

The Digital Invasion of the Teenage Brain

I opened an e-mail from a stranger one afternoon in May 2012. The message was from a young man who had recently read about my work on the adolescent brain, and in the subject heading of his e-mail he wrote, simply, "Computer Addiction." He began by telling me how at the age of fifteen he was lonely and introverted and spent most of his free time on his computer in teen chat rooms. It was easier to "meet" people this way and talk about his own interests anonymously, he said, and these chat rooms became a kind of obsession with him. The man writing to me said that he was now twenty-six and that over the years these online experiences had become more real—and more pleasurable—than his "off-line" experiences. "After that, my life went in a downward spiral," he said. He became "addicted" to chat rooms and increasingly felt as though his life was divided between his cyberself and his real self. Eleven years later this man felt confused and tortured about his computer addiction. He wrote to me, he said, looking for some perspective from me, and I hope I was able to give that to him. I told him being addicted to the

Internet involves the same reward center as drugs, and when he was a teenager, he was more susceptible to addiction in general, so it was understandable from a neurobiological perspective how he could get caught up in it. The digital world simply presented him with a means to interact with others at a time when that was enormously challenging for him, so it shouldn't make him feel guilty. After all, he did this self-searching totally on his own, without any guidance. The adolescent propensity for addiction occurs at a time of exploration when you're trying to make decisions but also, in the case of my correspondent, experimenting in a virtual world, so your perspective is skewed. He had no way of verifying what was real. Social isolation itself can be a stressor for teens who are roaming the digital world alone in their bedrooms.

Today's teenagers and twenty-somethings make up the first generation of young people exposed to such a breathtaking number of electronic distractions, and they are therefore susceptible to a whole new host of influences. Technology is another opportunity for novelty-seeking, and because the brain of a teenager is so easy to stimulate, all it takes is the latest digital toy to tease it into distraction. The cascade of neuroprocesses that kicks off the brain's reward circuitry and the rush of the pleasure chemical dopamine can be triggered just as easily by the release of the latest iPhone as by alcohol, pot, sex, or a fast car. In some ways, technology *is* a drug. Neither the American Psychological Association nor the American Psychiatric Association formally recognizes Internet addiction as a mental disorder, although the fifth edition of the *Diagnostic and Statistical Manual of Mental Disorders*, released in 2013, added Internet Gaming Disorder to its appendix and advised that additional study is needed. Both organizations, however, seem to be a bit behind the curve. There is increasing evidence of the effect of excessive Internet use on mood in adolescents, and several studies have shown a connection between depression, poor academic performance, and the inability to curb time spent online. In any case, increasing numbers of Internet "overusers" do, in fact, describe themselves as addicts and even seek professional help. In 2009, reSTART in rural Fall City, Washington, became the first residential treatment center in the United States specifically devoted to what has been termed "Internet addiction."

Today's teenagers are the world's leading authorities on technology, and while adolescents are the savviest of users, they are also the most vulnerable. Witness these headlines:

> *"Tech Addiction Symptoms Rife Among Students"*
> *"Students Are Addicted to Media Worldwide"*
> *"Technobsessed!"*

An experiment began in the spring of 2010 when two hundred students in a basic media literacy course at the University of Maryland were asked by their professor to do something unusual: go without their digital tools and toys—all media, in fact—for twenty-four hours. The results of the experiment, picked up by news outlets all over the world, prompted the professor, Susan Moeller, to conduct a second, much wider experiment. Both began with a simple request:

> Your assignment is to find a 24-hour period during which you can pledge to give up all media: no Internet, no newspapers or

magazines, no TV, no mobile phones, no iPod, no music, no movies, no Facebook, Playstation, video games, etc.

If you lapse by mistake (i.e. you answer a phone call without realizing it), do not then "give up." Note the mistake and go on to finish your 24 hours. If you do NOT make it the full 24 hours, be honest about it. How long did you make it? What happened? What do you think it means about you?

Although you may need to use your computer for homework or work, try to pick a time when you can go without using it—which may mean that you have to plan your work so that you can get it done before or after your 24-hour media-free period. You will not be judged on whether you went 24 hours, but we expect that you all will make it through the entire time without using any forms of media.

Moeller, who is a member of the International Center for Media and the Public Agenda (ICMPA) at the University of Maryland, partnered with the Salzburg Academy on Media and Global Change to conduct the second survey. They asked close to one thousand students in twelve countries, including the United States, to write about their experiences after their twenty-four-hour period of media abstinence was over, and when the students did, they poured out their angst:

"I began going crazy."
"I felt paralyzed—almost handicapped in my ability to live."
"I felt dead."

Across the globe, the same feelings were expressed again and again:

From the United Kingdom

"Emptiness. Emptiness overwhelms me."

"Unplugging . . . felt like turning off a life-support system."

"I feel paralyzed."

From China

"I sat in my bed and stared blankly. I had nothing to do."

"The feeling of nothing passed into my heart . . . I felt like I had lost something important."

From Uganda

"I felt like there was a problem with me."

"I counted down minute by minute and made sure I did not exceed even a single second more!"

"I felt so lonely."

From Mexico

"The anxiety continued for the rest of the day. Various scenarios came to my head, from kidnapping to extraterrestrial invasions."

From the United States

"I went into absolute panic mode."

"It felt as though I was being tortured."

Many of these students borrowed the language of substance abuse when they likened their media habit to an addiction and their self-imposed abstinence to drug and alcohol withdrawal. One US student wrote, "I was itching, like a crackhead, because I could not use my phone." A student in Mexico wrote, "It was quite late and the only thing going through my mind was: (voice of psychopath) 'I want Facebook.' 'I want Twitter.' 'I want YouTube.' 'I want TV.' " A college student in the UK wrote, "It's like some kind of disorder, an addiction. I became bulimic with my media; I starved myself for a full 15 hours and had a full on binge: Emails, texts, BBC iPlayer, 4oD, Facebook. I felt like there was no turning back now, it was pointless. I am addicted, I know it, I am not ashamed." Amusingly, the online media outlets whose headlines screamed addiction and warned about the all-consuming "technobsession" of the young provided multiple links, platforms, and interactive choices to "Follow" or "Share" or "Like" it on Facebook; to tweet it, "Get Alerts," and "Contribute to the story"; to send corrections, tips, photos, videos, or comments. No wonder that in trying to be media-free for a full day, many students also found themselves emotionally and psychologically distraught:

> *"I was edgy and irritated."*
> *"I got really anguished and anxious."*
> *"I was anxious, irritable and felt insecure."*
> *"I felt a strange anxiety."*

Moeller is neither a psychiatrist nor a neuroscientist, and her survey was more sociological than scientific. Still, it's hard not to read the responses of the experiments' subjects and wonder exactly what is going on inside the brains of young people who have been raised on digital technology. According to a 2011 study by the Pew Research Center's Internet and American Life Project, 95 percent of all young people, ages twelve to seventeen, use the Internet, and 80 percent use social media. Ninety-three percent have Facebook accounts, and 41 percent have multiple accounts. In an article written for a weekly teen publication, two Chicago high school students reported on the popularity of smartphones and the degree to which students will go to hide them from teachers and administrators. One student interviewed said, "Back in my sophomore year, I snuck my phone in as a biscuit sandwich in the morning. I covered it in [a] brown napkin and put it in between the biscuit buns. I would simply come to school and put my lovely cup of orange juice and tasty 'Bisquick biscuit' sandwich on top of the metal detector and walk right through." Another student said she would wrap her long hair into a bun before school and hide her phone inside. "Whenever the metal detectors beeped, they couldn't find my phone," she said. The level of attachment between teens and smartphones is so extreme, one high school senior told the authors of the teen publication article, "My phone has my whole life in it. If I ever lost it I think I would die."

So embedded in our consciousness are our smartphones that two-thirds of cell phone users report that they feel their phones vibrate when in fact they don't, a phenomenon researchers have taken to calling phantom-vibration syndrome. Judging from the above testimonials, it's not surprising to find that many of the same behaviors that typify the closet drug addict are also seen in Internet addicts: concealing behavior, lying, neglect of normal activities, and social isolation.

The compulsive need to be digitally connected happens on two levels, behaviorally and biochemically. Every ring, ping, beep, and burst of song from a smartphone results in an "Oh, wow" moment in the brain. When the new text message or post is opened, the discovery is like a digital gift; it releases a pleasurable rush of dopamine in the brain. In fact, there is mounting evidence that Internet addiction has much in common with substance addiction. Recent functional MRI studies in adolescents have shown that addiction to cocaine and meth alters connectivity patterns between the brain's two hemispheres as well as other important regions that use dopamine as a transmitter. What is interesting about the MRI studies of Internet addicts is that they are similar in pattern. Amazingly, unlike the effect on drug addicts, this neurobiological effect is not due to a chemical substance—it is purely a case of "mind over matter"! Hence, studies of Internet addiction may have revealed the purest circuits for addiction yet and may also prove to be good markers for rehabilitation in future treatment trials.

One of the most time-consuming Internet obsessions for young people is video gaming—electronic video games that involve human interaction with a user interface that generates visual and audio feedback. These games can be found everywhere—on computers, iPads, iPhones, Xboxes, Game Boys, you name it. Video games, of course, are more than half a century old. We can probably date the precise beginning to October 18, 1958, and, of all places, the Brookhaven National Laboratory on Long Island. The lab was holding its annual visitors' day, and nuclear physicist William Higinbotham, head of the instrumentation division, had an idea about how to provide a bit of educational entertainment. The result was the first real interactive electronic game: *Tennis for Two*. The game featured a two-dimensional "tennis court" on an oscilloscope, similar to an old black-and-white TV. The court was basically a vertical line down the middle and a brightly lit dot that left a trail as it bounced back and forth horizontally over the "net." Players served and volleyed from a console with buttons and rotating dials that controlled the angle of swing from an invisible tennis racquet. Would you be surprised to learn that even back in 1958 hundreds of people lined up to play the electronic game? No, I didn't think so.

Unlike a lot of the science in this book, what we know about the effects of video gaming on the brain comes almost exclusively from humans. After all, it would be pretty hard to simulate gaming in an experimental animal. Imagine a rat working a controller, playing a round of *Grand Theft Auto*! Not likely. So the effects have been measured either with psychological testing or with functional MRI studies. In the latter, researchers look at brain regions that get turned on and off in gamers and nongamers and measure the comparative size of brain regions. A study published in China in 2012 looked at seventeen adolescents who met the definition of gaming addiction and compared their brain scans with those of twenty-four nongamers of similar gender, age, and educational level. First, the gamers group scored much higher on tests for risk-taking. Next, functional MRI showed less connectivity to the gamers' frontal lobes, but more connectivity in areas that have been observed in nicotine addiction, for instance. These findings were also visible in a study aimed at measuring the actual thickness of the connections to the frontal lobe.

Another study, from Korea, corroborated the effect on brain structure in adolescents: fifteen adolescent males with the diagnosis of Internet addiction were compared with nongamers, and the finding showed the gamers had smaller orbitofrontal cortex regions, an area involved in modulation of risk-taking. This same pattern has been seen in people with obsessive-compulsive disorder.

Average young people, especially boys, will have played about ten thousand hours of video games by age twenty-one. This is a lot of time honing a skill that is not directly linked to any monetary or educational gain. In his book *Outliers*, Malcolm Gladwell says ten thousand hours is generally the amount of time required to become an expert in any field. This means that as a sideline, our youth are becoming experts in a skill set that has limited use outside itself, except of course for those who go into professions related to the gaming industry or whose job involves a lot of computer simulations. It also has been pointed out that ten thousand hours are more than it takes to get a bachelor's degree!

So is video gaming at a normal rate good or bad for the brain? The answer is not 100 percent clear. In a nutshell, it seems that a modest amount of gaming, like any form of learning, can actually be good for the brain. There is a difference between the hard-core gamers and casual gamers. Similar to reading and all other forms of "balanced" brain stimulation, developing superior skills at a video game has its upside. A study from the Max Planck Institute in Germany showed that gaming was associated with some regions of the brain being larger, in particular the entorhinal cortex, hippocampus, and occipital and parietal lobes. These are areas that are important for working memory and visuospatial skills. This sort of information is likely to be heartwarming for the many educators who are increasingly using video simulations that look a lot like games to teach many experiential skills, from flight schools teaching piloting to medical and nursing schools that simulate patients having heart attacks or strokes.

But obsessive gaming in the adolescent, to the exclusion of most other activities, appears, like addiction, to have both immediate negative effects and long-term negative effects on the brain.

Chinese researchers have discovered changes in the brains of college students who spend approximately ten hours a day, six days a week, playing online games. In these online gamers, the Chinese scientists found changes in small regions of gray matter responsible for everything from speech, memory, motor control, and emotion to goal direction and inhibition of impulsive and inappropriate behavior.

They also found that with increased time online the adolescent suffered increased shrinkage, sometimes as much as 20 percent. And there was more: when the researchers focused their scanners on white matter, they found abnormalities, specifically in white matter connectivity in the brain's memory centers, especially in the right parahippocampal gyrus. They hypothesized that an increase in density in white matter in this area of compulsive online gamers' brains could indicate problems in temporarily storing and retrieving information. A reduction of white matter in other nearby areas could impair the ability to make decisions, including the decision to turn off the computer or turn away from the online games! All these areas also have been implicated in alcohol, heroin, cocaine, and marijuana addictions in adolescents.

Perhaps the scariest digital temptations for teenagers are electronic games of chance and poker where they are suddenly vulnerable to a kind of double whammy addiction: gambling and technology. Various studies indicate that anywhere between 70 and 80 percent of all teenagers have tried online gambling at least once. Although you must be eighteen to place a bet in a casino, studies show kids as young as ten are logging on to Internet poker sites that offer free practice games. Subverting proof of age requirements on Internet gambling sites is made easier by the relative anonymity of, and 24-7 access to, these sites. There are online casinos located all over the world, and all who want to, including American teenagers, can play thousands of hands of poker every day as long as the sites they visit are based offshore where there are no age restrictions. Kids also can get started early on a path to addiction through free-to-play gambling apps available through iTunes.

The International Centre for Youth Gambling Problems reports that while 3 percent of adults are struggling with compulsive gambling problems, that number more than doubles to 8 percent when it comes to minors. And the temptations will only grow worse. According to a 2013 *Forbes* magazine article, Morgan Stanley predicts that by 2020 online gambling in the United States will produce the same amount of revenue as the Las Vegas and Atlantic City markets combined, or more than $9 billion.

Behavioral addictions are just as insidious as chemical addictions because they make use of the same brain circuits. This is why, whether it's gambling, interacting on social media, or snorting coke, teenagers are particularly susceptible to the rush of good feelings that comes with stimulating the brain's reward centers. CRC Health Group, the largest provider of specialized mental and behavioral health care services in the United States, believes there is such a thing as Internet addiction and on its website and in its literature lists both behavioral and physiological indicators:

Most nonschool hours are spent on the computer or playing video games
Falling asleep in school
Falling behind with assignments
Worsening grades
Lying about computer or video game use
Choosing to use the computer or play video games, rather than see friends
Dropping out of other social groups (clubs or sports)
Being irritable when not playing a video game or not being on the computer

Carpal tunnel syndrome—joint pain in fingers, hands, and wrists—a consequence of repetitive motions that come with excessive keyboard use

Insomnia

Forgoing food in order to remain online

Neglecting personal hygiene and grooming in order to remain online

Headaches, back pain, and neck pain

Dry eyes and vision problems

Addiction, of course, may not be the only hazard of Internet obsession. A 2006 study reported in the *Annals of General Psychiatry* looked at the link between video games and symptoms of ADHD in adolescents and found that more symptoms, and more *severe* symptoms, of ADHD and inattention were found in adolescents who played video games for just an hour or more a day.

Which brings us to the topic of multitasking. While there is increasing evidence that adolescents are more vulnerable to Internet addiction, there are mixed opinions as to whether the digital invasion of our environment impairs an adolescent's ability to focus the way adults claim it does. Can teenagers really multitask better than adults? After all, we know adolescents have a heightened ability to learn during their teen years, so perhaps they can. Yes and no.

When asked about multitasking, most teens say they believe they are good at it and that it allows them to accomplish more. On the other hand, studies show that multitasking actually interferes with learning in adolescents and that it takes anywhere between 25 percent and 400 percent longer for a teenager to complete his or her homework if multitasking is involved. So why do teens profess that multitasking helps them? It may be because multitasking makes them feel emotionally satisfied. For example, in one survey researchers found that students who watch television while reading report feeling more satisfied than those who read without watching TV. Zheng Wang, the lead author of the study, explained it this way, saying, "They felt satisfied not because they were effective at studying, but because the addition of TV made the studying entertaining. The combination of the activities accounts for the good feelings obtained."

Remember the Minnesota undergrads who showed that distractions during memorizing and test taking lowered their scores? Not only is multitasking an impediment to learning, say scientists, it also can prompt the release of stress hormones such as cortisol and adrenaline. Chronically high levels of cortisol have been associated with increased aggression and impulsivity, loss of short-term memory, and even cardiovascular disease. In other words, multitasking can wear us down, causing confusion, fatigue, and inflexibility. We continue to do it in large part because of habit, and habits for adolescents are particularly difficult to break; that is why as teens get used to multitasking, they are more likely to continue doing it. "This is worrisome," Dr. Wang of Ohio State told the media, "because students begin to feel like they need to have the TV on or they need to continually check their text messages or computer while they do their homework. It's not helping them, but they get an emotional reward that keeps them doing it. . . . If you multitask today, you're likely to do so again tomorrow, further strengthening the behavior over time."

Despite the emotional rewards teens seem to get from multitasking, some researchers have found a correlation between multitasking and symptoms of depression and anxiety. At this point, however, scientists don't know if increased multitasking leads to those symptoms or whether those symptoms lead to an increase in multitasking. The best way to help teenagers avoid the temptation to multitask is to encourage prioritization and structure. Encourage your adolescents to make lists—such as what they need to take home from school in the afternoon in order to do homework, or what they need to accomplish before going to bed. Try to get them in the habit of crossing these things off a list, too, as they are achieved. When your children come home from school, make them clean out their book bags or knapsacks in front of you and organize their homework assignments, and then ask them which they need to do first. Your teenagers may do this kicking and screaming, but if *you* make it a priority—no TV, no computer time, no snack until certain things are done—then you'll increase your chances of success. Normally the fewer distractions the better, which is why you want to make sure the TV is turned off in the background when your son or daughter is doing homework. Some teens, of course, might actually relax and concentrate better listening to music on headphones while they do their homework. And the only way to be sure is to observe them.

The ramifications of adolescent involvement, or overinvolvement, with technology can affect a person not only cognitively and emotionally but also legally. In January 2013 an eighteen-year-old Oregon man, Jacob Cox-Brown, posted the following status update on his Facebook page: "Drivin drunk . . . classsic ;) but to whoever's vehicle i hit i am sorry. :P." The confession wasn't sufficient to warrant a charge of drunk driving, but when the local police became aware of the post, they showed up at Cox-Brown's door and arrested him anyway, charging him with two counts of failing to perform the duties of a driver. Six months earlier, an eighteen-year-old Kentucky woman posted a message after being arrested for drunk driving and hitting another car. In her Facebook post about the incident she added the ubiquitous abbreviation "LOL" (for "Laugh Out Loud"). Not taking kindly to the seemingly flippant message, the judge jailed her for forty-eight hours. States have varying laws regarding texting and driving; some even prohibit teenagers from any use of a digital device while driving, including talking on a cell phone, even if it's with a hands-free device.

Another consequence of the Internet is that it brings a breadth of stimulation of all kinds into a person's intimate environment, allowing teens to be exposed to dozens of experiences a day, far more than previous generations had. The converse is also true: that the actions of a teen can resonate through a much larger community than in the past. A teen prank, which in days of yore would have been confined to the school yard, can go viral with countless unintended consequences. I have secondhand knowledge of this. A colleague of mine is a single parent with a sixteen-year-old daughter who is a sophomore at a public high school in Philadelphia. Like all her peers, the girl has a smartphone and spends much of her free time on the Internet, texting and tweeting. Another student at her high school took a surreptitious photograph of the daughter of this colleague. The photo shows the girl with her head bent down and her eyes closed, ostensibly asleep in class. The female student who took the photo posted it on Instagram with a derogatory caption. It didn't take long for the young girl to see the picture and caption online, and after she called home, upset, her mother called the high school. By the end of the school day, the student who snapped the photo and posted it on Instagram was suspended. She was also angry and looking for revenge.

Because the daughter of this colleague liked to tweet about where she was after school and what she was doing, she wasn't hard to find. The girl who had been suspended tweeted out that she was going to "beat up" the girl she'd taken the picture of, and invited others to come to Center City in Philadelphia to watch. Dozens did, and what ensued was a near riot in which four adults and ten teenagers were arrested and the photographed girl, who had been assaulted, suffered a few scratches and bruises. When her mother told me the story, she said, "I hate social media for teens. They can't handle it. It's too much freedom. They talk about whatever—the foul language. It's a public arena to say or do anything." When I asked about the effect on her daughter, she said that her grades went down for a while as she struggled with the embarrassment at school, but she rebounded and "said she really learned a lot."

"I hope and pray she did," her mother added.

The consequences of misuse of digital media can be far more severe than a couple of days in jail or a public fight. Tyler Clementi, a shy eighteen-year-old violinist from Ridgewood, New Jersey, was barely a month into his freshman year at Rutgers University when he was caught in a cyberscandal not of his own making. The slender redhead, who weeks earlier had told his parents he was gay, had recently been assigned a roommate, Dharun Ravi, a self-professed computer geek. On September 19, 2010, Ravi, while absent from the room, surreptitiously used his laptop webcam to spy on Clementi's intimate encounter with another man. In a nearby dorm room belonging to freshman Molly Wei, Ravi used Wei's computer to connect to his laptop, logged on to the website iChat, and activated the webcam back in his room. For a minute or so Ravi, Wei, and several others watched Clementi and his male friend embrace and kiss. The next day Clementi found out about the webcam because Ravi tweeted about it. Clementi seemed to take the betrayal in stride, and the following day simply requested a room change. Two days after the incident, however, Clementi discovered Ravi was poised to spy on him again.

On September 22, at about 6:30 p.m., Clementi boarded a university bus in New Brunswick, New Jersey, then took a train into New York City. At 8:42, on that warm, rainy first day of autumn, the young gay man, who had recently been awarded a prized seat in the university's orchestra, posted a final status update on his Facebook page: "Jumping off the gw bridge sorry." It's not clear if Ravi was aware of the message, but five minutes after Clementi posted it, Ravi texted his roommate to apologize: "I'm sorry if you heard something distorted and disturbing but I assure you all my actions were good natured." The following day police discovered Clementi's body floating in the chilly Hudson River below the George Washington Bridge. Six days later, Ravi and Wei were both charged by the Middlesex County prosecutor's office with invasion of privacy.

The suicide and webcam scandal made headlines around the world, from England and France and Denmark to Turkey, Japan, Indonesia, and Australia. Celebrities, politicians, and talk-show hosts called the spying incident cyberbullying, a hate crime, and worse. Dueling Facebook pages popped up condemning Ravi and Wei, and supporting them. The two eighteen-year-olds received death threats, were forced into hiding, and finally, under withering public scrutiny and scorn, withdrew from school.

Eventually Ravi was charged with bias intimidation, witness tampering, and evidence tampering. Wei accepted a plea deal and was given three hundred hours of community service. Ravi, proclaiming his innocence through his lawyer and saying the incident was a stupid prank and not an act of bias, turned down two plea offers and went to trial in February 2012. After three weeks of witnesses and without ever testifying himself or addressing the court at sentencing, Ravi was found guilty on fifteen charges, including bias and invasion of privacy. Facing a possible ten-year prison term, he was, instead, sentenced to thirty days in jail, six hundred hours of community service, and probation. Many believed that he deserved a stiffer penalty, others that thirty days was too much. Regardless, his life and Wei's were irrevocably altered and Clementi's was cut all too short. Few believed Ravi's act was motivated by bias; most believed he was either showing off for his friends or simply being a virtual voyeur. And no one knows the extent of Clementi's anguish at being gay and whether there were difficulties with his family's acceptance of his sexuality.

Nearly every day there are news reports of cyberbullying, digital invasions of privacy, and Internet communications gone horribly awry. Many, if not most, involve teenagers. In 2008 in Cincinnati eighteen-year-old Jessica Logan hanged herself after an ex-boyfriend forwarded her nude cell photos to high school classmates. In 2006, an eighth grader in Missouri killed herself when she learned an Internet romance was a hoax. And in 2001 an Oregon State University engineering student was convicted of invasion of privacy for using his laptop webcam to broadcast over the Internet images of his roommate and his roommate's girlfriend having sex. Teenagers have always committed careless, impulsive acts, but the digital tools now at their disposal have exponentially magnified the dangers and certainly the consequences of those careless, impulsive acts.

For Ravi, a teenager who regarded himself as a computer expert, those consequences were never seriously considered until after the fact. For Clementi, seeing beyond the incident or finding help for the overwhelming despair that swept over him in the hours before he jumped to his death was obviously impossible.

There is no turning back from the digital world we all live in, but we can turn away—if even for a few hours or minutes a day—and the earlier we start doing this with our kids, the better. Limiting a teenager's use of the Internet isn't easy, but one way to better control it is moving the computer out of your high schooler's bedroom and into a common area where you can check more readily on what your son or daughter is up to. Software programs can help you monitor what sites your kids visit and block access to others, but the main responsibility is for you to communicate with your teenagers. Familiarize yourself with what they do online and what sites tempt them the most and when—for instance, during math homework or when they're supposed to be getting ready for bed. Try to approach the problem not as something your teen is being punished for but as something he or she needs help with in order to stay balanced, well rounded, and less isolated.

Believe it or not, even some tech executives are beginning to realize that digital accessibility may not always be a good thing. In 2012 the *New York Times* reported a number of Silicon Valley executives admitted not only to digital overload but also to the need to take time away from technology. Stuart Crabb, a director at Facebook, gave the *Times* the following analogy: "If you put a frog in cold water and slowly turn up the heat, it'll boil to death." Crabb said it was important for everyone to be aware of how time spent online affects not only job performance but also relationships and overall quality of life. How serious are these digital trailblazers about heeding their own advice? The chief of technology at Cisco, Padmasree Warrior, told the *Times* that she regularly advises the 22,000 employees under her to disconnect and take a deep breath. She does, she says, every night when she meditates and every Saturday when she paints and writes poetry. Her cell phone? She simply turns it off.

14
Gender Matters

The first talk I ever gave on the teenage brain was in 2007 at my sons' school, Concord Academy. The principal, guidance counselor, and teachers had been so good to my kids and really helped me get them through a few chaotic teenage years, so I wanted to give something back. An idea that began as a single lecture turned into a two-day symposium with sessions for teachers, parents, and kids. Two friends and colleagues filled out the roster of speakers: David Urion, MD, an associate professor of neurology at Harvard Medical School who also treats children with cognitive impairments, including autism and learning disabilities; and Maryanne Wolf, the director of the Center for Reading and Language Research at Tufts University.

David is a neurologist and an expert in ADHD and, among other things, has spent a lot of time trying to understand the effect of learning disabilities on children and adolescents. Maryanne's research and writing delve deeply into how children learn to read and the differences between boys and girls in language processing.

When it was Maryanne's turn to talk to the students at the symposium, she opened with a quick demonstration of one of those differences. First she asked for two volunteers in the audience, a boy and a girl, both thirteen. Then she told them she was going to ask each of them to name as many words beginning with a certain letter as they could, given a time limit of one minute. The same thing happens every time she does this exercise with a teen audience. She always lets the girl go first, giving the boy even more time to get the hang of things. So the thirteen-year-old female volunteer stands up, and Maryanne says, "Name as many words as you can that begin with the letter *P*." And off she goes—"pumpkin," "pattern," "public," "popular," and so on. By the time a minute is up the girl has rattled off thirty-five words. All this time the boy has been watching and preparing, right? So then Maryanne turns to him and says, "Okay, are you ready?" A few of the boys in the audience chuckle and snicker. Then Maryanne says something like, "Okay, the letter is *M*." Immediately the boy starts off by looking around, as if he's actually looking for cues for words, and he begins to struggle, hemming and hawing, and by the time a minute is up he's lucky if he has named half as many words as the girl. I've seen Maryanne do this now on three separate occasions, and always with the same result. There is a reason, she explained, that the girl did so much better than the boy and that in a couple of years that difference will be negated. The ability to fire off the words actually relies on two distinct brain areas: the parietotemporal area, where speech and language are processed; and the frontal lobe, which controls decision-making. The task the two teens were asked to perform requires both language and rapid decision-making, and at the age of thirteen, girls are simply further along in having those two required brain areas wired together.

Scientists and psychologists have long known that there are differences in development between girls and boys and that girls' language development, specifically reading and writing, is generally about one to one and a half years ahead of boys'. As parents of teenagers you all are probably nodding your heads right now. I'm not telling you anything you don't already know. What you may not know is that these differences cannot all be laid at the feet of a difference in the speed of development. Why? Because there truly are anatomical, physiological differences between an adolescent girl's brain and an adolescent boy's.

Most of the variations in brain structure between males and females are minimal and relative to the difference in average body size between men and women. Others don't correlate with any specific advantage or disadvantage. For instance, adult male brains are on average 6 to 10 percent larger than female brains, but there is data from Harvard researchers to suggest females have more connectivity between hemispheres. The differences can be even more exaggerated in childhood, when boys and girls of the same age can have as much as a 50 percent difference in brain volume during the steep part of the growth curve. All of this makes it difficult, and indeed foolish, to draw conclusions about differences in brain function based on differences in anatomy, at least when it comes to talking about males and females.

The fact that there *are* differences in neural anatomy between the two sexes, however, is undisputed. The differences are present in early fetal life, as hormones already have altered the destiny of brain regions that are set up to go either way in the embryo. This is called sexual dimorphism, and one region that is heavily altered by early differences in levels of the female hormone estrogen or the male hormone testosterone is the hypothalamus. This turns out to be very important because the job of the hypothalamus throughout life is to regulate hormones in women and men.

Sandra Witelson is a professor of psychiatry and neuroscience at McMaster University in Ontario, Canada, and owns the world's largest collection of normal brains, with a total of more than 120. In thirty years of research she has consistently found differences between male and female brains, but the differences are subtle and the relationship of those differences to function is often unexpected. For instance, the size of the corpus callosum, the strip of neural tissue that links the left and right hemispheres of the brain, appears to be linked to verbal ability on IQ tests in women but not in men. (And in adolescence a girl's corpus callosum is about 25 percent larger than a boy's.) In another finding, memory in males was correlated with how closely the neurons were packed together in the hippocampus, but this was not true of females. Studies involving cognitive tasks do not show gender-based increases or decreases. What all this research underscores is the need to be intellectually open when it comes to making gender-specific delineations between male and female brains.

More recently, Raquel Gur and her colleagues at the University of Pennsylvania have used a combination of MRI techniques—diffusion tensor imaging (DTI), which maps physical connections in the brain; and fMRI, which maps how synchronized brain regions are when they activate—to examine the connectivity between brain areas in women's versus men's brains. When Gur and her colleagues saw one brain area turn on or activate another, they drew lines between them, mapping the "connectome" of the brain. They compared the amounts of connectivity between the two hemispheres and found that, while

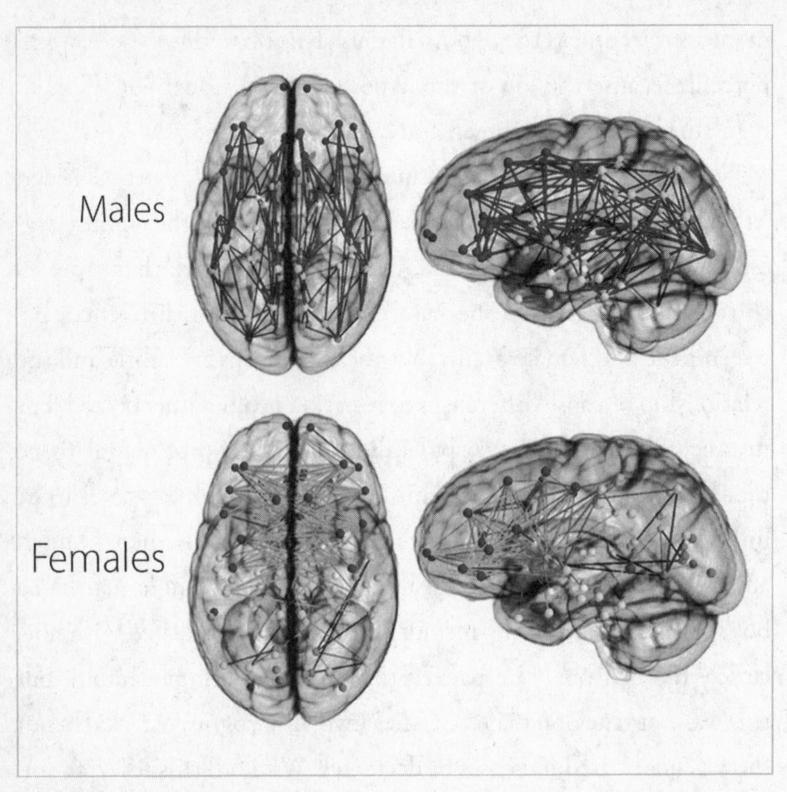

FIGURE 28. Gender Differences in Brain Connections: Both functional and diffusion tensor magnetic resonance imaging can be used to look at how brains are organized and connected. Females appear to have more connections between hemispheres than males, but males appear to have stronger connections within each hemisphere.

the vast majority of connections were the same between men and women, men have more connections within hemispheres, while women have greater connectivity between hemispheres.

At the same time, it's indisputably true that at least in adolescence there are real differences in certain brain functions between males and females. Because of their larger corpus callosum, which means better communication between the brain's two hemispheres, girls may have a greater ability to switch between tasks than boys. A friend of mine, for instance, who also has two sons—one still in his teens, the other just barely out of them—recently threw up her hands in that classic parental gesture of frustration and surrender. My friend's daughter was about to get married somewhere in the Midwest. She'd planned the wedding and set the date a year in advance. Everyone was flying in for the event, so of course everyone needed a government-issued photo ID to get on the plane. But a few days before the event the bride's brothers were sent scrambling. My friend's younger son was staying in Vermont at the time, and he suddenly realized his driver's license had expired and his passport was at his father's house in Massachusetts. His older brother, who was living in Washington, DC, lost his wallet in a taxi, and he, too, had left his backup ID, his passport, in Massachusetts. But at least *he* notified his father of his problem a few days in advance—enough time, or so we thought, for his father to express-mail the passport. Alas, the older boy claimed he never received the package and so still couldn't get on the plane. In fact, the only reason he eventually did get on was that he got a signed affidavit from his employer verifying his identity! Both boys made it in time for their sister's wedding—barely!

We know from earlier chapters that both males and females lose gray matter between the ages of six and eighteen and both gain white matter throughout adolescence and well into their twenties. While boys gain white matter faster than girls, they are not necessarily using the same parts of their brains when they perform the same cognitive tasks. In 2008, researchers at Northwestern University and the University of Haifa in Israel collaborated on a study of how boys and girls process language. In general, adolescent females have superior language abilities compared with adolescent males. When the scientists gave their female subjects complex auditory and visual language tasks, the activated areas of their brains were associated with abstract thinking through language, and their level of ability was correlated with degree of activation. The same correlation held true for boys, but their accuracy depended not on the abstract language areas but on their senses of hearing and sight. There have been several studies that show that male and female adults use different parts of the brain to sound out words or read aloud: different paths can lead to the same result.

The amygdala, where emotions generally arise, develops about eighteen months sooner in girls than in boys in early adolescence. The hippocampus also develops earlier, and there are differences between males and females, with the two sides of the hippocampus being asymmetrical in men and symmetrical in women. This is consistent with other data showing higher levels of side-to-side connectivity in females in general. Both the amygdala and the hippocampus are in the limbic system, and their function can be affected by hormones.

How many times did I want to pull my hair out when my boys would mope around the house and have nothing to say at mealtime, all because they'd been left off a party list, lost a sports competition, or broken up with a girl? It was like extracting teeth to get them to tell me what they were thinking and feeling. I envied some of my friends whose teenage daughters, while firmly seated in that emotional roller coaster called adolescence, still could open up to their mothers or fathers about their feelings.

Both boys and girls show large swings in their emotional behavior during adolescence. In part, they are experiencing for the first time the effects of hormones, not yet having learned how to control them. This control will eventually involve the frontal lobes, which will dampen the swings, but this area, we know, is not yet fully available to them. They are on a steep learning curve! A teenager faces double trouble in terms of trying to process certain emotional experiences, especially when the brain structure responsible for integrating emotional information with memory is still under development. Teens react more or less instinctually to the events around them precisely because those connections between the emotional and intellectual parts of the brain, including memories of similar events in the past, are still being formed. But in this regard, girls may have a slight edge over boys, at least in early adolescence.

When we think about gender differences, we often think about emotion. However, there are other ways that the different rates of brain development manifest themselves. An obvious one is organizational skills. Organization requires brain connectivity and integration, not just raw intelligence and synaptic power. Myelination plays a huge part in this, and as we have said earlier, it requires the better part of the first three decades of life to be fully completed. The time of greatest gender disparity in this process occurs during adolescence.

Many learning specialists will testify that boys take longer to develop their organizational and attention skills, and the practical implications for educators can be profound. A good friend of mine is an educational consultant, and her job is to place students in private schools and colleges. For many teenagers, especially boys, this can be a difficult process. The steps required to gain acceptance to a good school or college are complex. Thirty years ago, an application could be submitted to Harvard, UCLA, or NYU without much handwringing; today, by contrast, the steps to enter such schools are numerous and the competition is steep. For boys, who lag behind girls in terms of organizational skills, the process is that much harder.

My friend the consultant told me the story of a sixteen-year-old boy named Ryan she'd recently been counseling. He was a star hockey goalie on his high school team, and at the start of his junior year coaches from several excellent private colleges in the Northeast contacted him and asked him to apply. Naturally, they requested his grades and test scores. Ryan, apparently, is an intelligent and highly motivated student. However his grades, in the C+ to B range, were probably too low for acceptance to these schools, even with a coach's endorsement. When the consultant asked him about his grades, he complained about the "three hours of homework every night" and his many other commitments, such as volunteer work and varsity sports, which seemed to make it harder for him to succeed at academics. His mother told her, "Truth is, Ryan has trouble getting his homework in. He's disorganized and waits to the last minute to study for tests; then he'll spend too much time on his history, for example, and completely forget the math test."

Ryan's academic approach is very common for a teenage boy because a rigorous high school curriculum these days requires superb attention, planning, and organizational skills, all of which develop more slowly in boys. This hit home for me with Andrew, who was challenged early in high school to figure out how to be better organized in order to get his work done. It took him a year to turn things around, and during that time I kept in close communication with his high school guidance counselor. Andrew knew it was up to him to take responsibility for his homework, for his sleep habits, and for the distractions that kept him from studying. With gentle nudges at school and at home, oversight, and assistance, he not only became more disciplined about how and when and where he studied but also became more confident.

I was reminded how different it was up until the last decade in England, the country of my parents' birth, where all students had to take a common entrance exam at eleven years of age; if they didn't do well they couldn't go on to A levels, and that meant they could not go to college. When I was young, education wasn't a right as much as a privilege in the United Kingdom, and because not everyone got to attend a university, education became one more caste system. It's truly a shame that in so many other countries, before a child even reaches puberty, he or she has already been tested, evaluated, and judged to be either intellectually worthy or unworthy of higher education. If my sons at ages eleven or twelve or even fifteen or sixteen had been subject to this kind of life-determining "tracking," I'm not sure they would be the highly successful college-educated men they are today. There is so much on the line for our teenagers, it seems incomprehensible that their futures should rest on an evaluation of their not yet fully developed brains.

Given the current statistics—that more girls have higher average SAT scores than boys and that girls are more likely to complete high school and to enroll in both undergraduate and graduate educational programs—things are definitely changing from prior generations. Still, stereotypes are sometimes hard to break. Break them, however, we do.

What stereotyping confirms is that our ideas about gender differences are almost always behind the times—and behind science, too. Even today, many people will cite apparently scientific evidence for how good men are at visualizing space; they're logical and linear. Women are good at intuiting; they are more creative and empathetic and see things holistically. My personal observation is that these black-and-white stereotypes are inaccurate, and there is an ever-increasing amount of data to back that up. In middle school, for instance, 74 percent of girls express interest in science, technology, engineering, and math (STEM), according to Girls Who Code, a national nonprofit organization that was launched in the spring of 2012 and seeks to close the gender gap in science, engineering, and technology. Girls Who Code is supported in part by Twitter, Google, General Electric, and AT&T, and fosters programs that "educate, inspire, and equip high school girls with the skills and resources to pursue computing careers." The problem with the numbers of girls interested in these subjects, says the organization, is that they peak in middle school. By the time it comes to choosing a college major, only 0.3 percent of high school girls select computer science. Janet Hyde, a researcher at the University of Wisconsin–Madison, found that girls who grow up believing boys are better at math—something parents and teachers persist in thinking—are more likely to avoid the harder math courses. Consequently, this can result in a self-fulfilling prophecy and may contribute to the gender gap at the highest levels of math achievement, which are reached disproportionately by men.

The numbers, however, have nothing to do with aptitude, especially now. Hyde also looked at annual math tests that were required by the No Child Left Behind law in 2002. Twenty-five years ago, girls and boys did equally well in elementary school, but boys far outstripped girls by the time they reached high school. In their recent study of more than seven million children's test scores in ten states, Hyde and her team found no difference between boys and girls in either middle school *or* high school. For today's teenagers, it seems that the field is starting to equalize in terms of education, making it all the more important that we get gender-based differences in adolescent learning right.

The differences are not substantial enough to preclude children of either sex from doing and becoming whatever they want in life. However, for me the differences do raise some questions as to whether in certain circumstances we should consider *some* gender-based high school curricula. Given the couple of years' disparity between peak cortical volume in girls versus boys, one might predict that girls, who reach specific levels of cognitive development before boys, could benefit more from math and science courses at earlier ages.

Ultimately, differences in the time, size, and rate of brain change between males and females and the effects of sex steroids are too complex to let us make conclusions about variations in brain function between the sexes. Despite all the books and articles and TV programs detailing the differences between a "pink" brain and a "blue" brain, no causal link has ever been established between gender-based variations in brain development and the cognitive abilities of females versus males.

What scientists do know is that the brain, at any time in its life but especially during adolescence, is a product of both nature and nurture, including all the exposures, stresses, and stimulations of a person's environment. One important observation is that in the developing world, puberty is being reached at ever-earlier ages, and the theorized causes of this have ranged from environmental effects to advanced nutrition to steroids in our food supply, but the jury is still out. What relevance this has to brain maturation is very unclear and likely to be a subject put under the microscope in the coming decade. While people can argue about toxicity and dangers of certain environmental chemicals, what is incontrovertible is this: What we learn and experience, the good and the bad, the mild to the severe, will change our brains.

15
Sports and Concussions

I was at home, working, on a Saturday morning in January 2010, when a message arrived in my e-mail in-box with these words in the subject heading:

> Help with 15 year old's seizures no one can diagnosis it the past three weeks have been very scary. Please read this information.

I then read two pages of notes, single-spaced, from a desperate New Hampshire woman looking for help for her once healthy teenage daughter—and the situation was indeed scary. Maureen's daughter, Holly, was about to turn sixteen. Two years earlier, Holly had been an avid athlete, a competitive cheerleader, a track star, a snowboarder, and a member of the boys' football team. Her only issue was an anxiety disorder, but it barely seemed to slow her active extracurricular life. Then, in October 2007, Holly suffered two concussions in the space of just two weeks, both of them from football collisions.

The first was in practice; the second was courtesy of an opponent who Maureen believed had targeted Holly because she was a girl. For the next few months her daughter seemed fine, but in March 2008 a series of events set off alarm bells. It began with a constant headache, which Holly had for two weeks. Then, sitting at the computer at home, she suddenly passed out, without warning, and smashed her face into the computer's keyboard. Holly was so scared she ran to her parents, who immediately took her to the local hospital for an EEG of her brain. The diagnosis: "Teenage girls just pass out." Three days later, however, there was an escalation of symptoms. Holly was at a friend's house when she suffered a seizure: her eyes rolled back and fluttered; she drooled and again passed out. Again she was rushed to the ER and given an EEG, and again came the same diagnosis: "Teenage girls just pass out."

The following day, her parents took her to their primary care physician, who evaluated her for either basilar migraine, a migraine with an aura that originates in the brainstem, or epilepsy. A few days later, Holly had a grand mal epileptic seizure at home that lasted more than four minutes. Her doctor immediately put her on anticonvulsant medication, and for months she was symptom free. She even started playing football again, but twice in practice she reported having chest pains and shortness of breath. Even though a cardiologist could find nothing wrong, she eventually had to quit the team. When Holly started the ninth grade, in 2009, she hadn't had a seizure in three months. But in January 2010, working on a report in the school library, she had a seizure that lasted more than six minutes. Two weeks later she had another, in French class, which lasted for eleven minutes. And so it went. Though her first seizure didn't happen until months after her injury, the seizures had gradually increased in number and intensity until Holly's mother was at wit's end. I recommended a neurologist in her area.

Head trauma can have lots of consequences, even when there is no obvious major brain injury. There are many new studies that have focused on early and late effects of concussion, and many are showing that adolescents' brains may take a kind of "hit" that is different from adults'. In some cases, for an injury of given severity, an adolescent fares worse than an adult. This is an increasing issue because physical contact has been progressively escalating in high school and middle school sports for both sexes, partly because of Title IX. In addition, there has a been a lot of controversy around closed-head injuries in the military from IED blasts, and the fact that the young service people affected are often still in their teens or early twenties, with not yet fully mature brains, is cause for concern.

Holly's case was clearly unusual, but it was also deeply troubling because more girls participate in athletics than ever before, and even in traditionally noncontact sports like soccer and field hockey they are suffering concussions, whether from heading a ball or being knocked to the ground, at a rate far greater than boys. Only 5 percent of closed-head injuries, which include concussion, result in the development of a seizure disorder like Holly's. She suffered the unfortunate consequence of developing epilepsy, something that happens in about 20 to 30 percent of people with a severe head injury. When epilepsy is acquired, it is more commonly the result of a severe, and immediately life-threatening, injury, either a compound skull fracture or a penetrating brain wound. Post-traumatic epilepsy is rarely the result of an athletic injury. But Holly, unbeknownst to her family, had three strikes against her: she was an adolescent, she was female, and she'd had several concussions within a short period of time.

Concussions have gotten increased attention in recent years, but mostly as they relate to men and professional sports. According to the American Academy of Pediatrics, however, the second highest rate of sports-related concussions in high school after boys' football is girls' soccer. Concussions can happen in any sport. I had a patient who was a high school wrestler, a top student, when I saw him one September to evaluate him after his second concussion, which he received at wrestling camp over the summer. He came to me because he was having headaches and attention and memory problems. He had scored very high on his practice SATs the prior spring, and was gearing up to take the real SAT that fall. With the second concussion he lost consciousness momentarily and was rushed to the local ER, where a head CT scan showed no visible damage. He was sent home and told to rest for a couple of weeks, and during that time he had problems sleeping and frequent headaches with some nausea and irritability. When he started back at school, on the first day of classes, things did not feel right. His mom was distraught because her A student son could not focus on his schoolwork and could not even complete a simple assignment. He was moody and agitated, too. When I saw him, his neurological physical exam was really quite normal, but he did falter on some short-term memory and attention tests. I had to make sure he had not developed epilepsy, as Holly had. Often a few, or even more, months go by before recurrent seizures are noted. Reassuringly, his EEG was normal. As is typical for postconcussion patients, though, his headaches had a migraine quality, with light bothering his eyes when the headaches were especially severe. As they were occurring daily, we put him on an antimigraine preventive medication, and over a few weeks his headaches decreased in severity and frequency.

In the meantime, however, he became increasingly anxious about his grades and also about getting another headache, and the anxiety eclipsed the other problems and became disabling. He withdrew from family and friends and continued to be moody. Furthermore, when he took the SAT, he sorely underperformed; this further increased his anxiety. The SAT score did confirm that his concussions had caused a quantifiable drop in cognitive performance, so he was now on everyone's radar screen. Neuropsych testing was done, and it documented a learning deficit and inattention. These were totally new problems for this teen and ones he might have to battle for a long time.

This is just one case, and hardly an isolated one, of what repeated concussions during adolescence can do to mental and scholastic performance. But there are also myths about concussions I want to dispel right off the bat. First of all, concussions do not affect everyone the same way, and some people may even have a genetic predisposition to them. Also, concussions are a problem not just in contact sports but elsewhere; they can occur in noncontact sports, in car accidents, falls, and even from a severe jostling. The medical community now realizes that concussions can happen without a clear episode of blackout.

More than a decade ago the American Medical Association linked sports-related concussions to lower scores on several tests of cognitive function, but only recently has research helped us realize the complex and frightening truth about the dangers of these concussions on still-maturing brains. Among high school sports with male/female participation—soccer, lacrosse, basketball, baseball, softball, and gymnastics—girls sustain concussions nearly 70 percent more often than boys, even though boys participate in these sports at a rate slightly higher than girls. Moreover, in soccer in particular, female concussion rates are three times the rate for male players. In a study based on a survey of more than four hundred high school athletic trainers, researchers also found it took girls substantially longer than boys to recover from their symptoms and return to action. After a concussion, adolescent girls score significantly worse on visual memory tasks than boys and show greater reductions in reaction times on mental tests.

In order to understand why adolescents, and in particular adolescent girls, are so susceptible to these injuries, we have to understand exactly what a concussion is. That hasn't always been easy because a concussion is by definition a closed-head injury. In other words, it's an injury to the brain where there is no, or very little, indication of a wound to the head or skull. Essentially, the brain is a soft organ that floats in a protective sea of cerebrospinal fluid inside the skull. The fluid acts as a kind of cushion, protecting the brain from normal jostling of the head. With a concussion, however, that cushion isn't thick enough to absorb the blow as the head is violently snapped forward and backward, in whiplike fashion. The injury occurs when the force is so great that it causes the brain to hit the side of the skull, damaging neurons. A hit so violent that the brain bounces off one side of the skull, only to rebound and strike the opposite side, is called a coup-contrecoup concussion.

The kind of power generated in athletic collisions can be measured in g-forces. A g-force is a measurement of acceleration in a cause-and-effect situation, and it is proportional to the reaction force experienced by the object or person undergoing the acceleration. A sneeze, for instance, generates just under 3 gs on the body, mostly the head; a slap on the back generates a bit more than 4 gs; and plopping down in a chair, 10 gs. In a low-impact rear-end collision, if the trailing car is traveling about ten miles an hour, it will hit the car in front of it with an impact in the 10-to-20 g-force range. If the impact is between 20 and 30 gs, you're in a pretty bad car accident. A hard hit in the NFL can reach into the 30-to-60 g-force range. And an impact that produces 90 to 100 gs—the force of smashing your head against a wall at twenty miles per hour—will usually cause a concussion.

Football players routinely hit one another with forces in excess of 100 gs, and 150-g hits are not unheard of. In fact, Purdue researchers recently evaluated a high school football player they estimated to have received a blow to the head that carried the force of 289 gs—that's nearly a hundred times more than the sustained g-forces associated with a shuttle launch—and yet this high school player had no outward signs of a concussion, nor did he report any symptoms. And therein lies the problem. Over the past few years, scientists have slowly begun to realize that brain damage can result even from nonconcussive blows to the head. All it takes are repetitive strikes of moderate intensity. In other words, thousands of kids playing contact sports who have never had to sit out a game because of a concussion could be at risk of brain damage—brain damage that is going undetected and undiagnosed and will be likely to cause cognitive impairment later in life. Those Purdue researchers who evaluated that "nonconcussed" football player who had absorbed nearly 300 gs discovered this frightening fact only by accident, after the young man volunteered for the study.

Several years ago Eric Nauman, a biomedical engineering professor, and a handful of colleagues were studying concussions among high school football players. In order to do a proper analysis of the brain changes of the concussed players in the study, they needed a control group with which to compare them—that is, high school football players who had never been diagnosed with a concussion. So early in the football season the researchers recruited a number of these players from local high school teams and scanned their brains one by one. When Nauman and his colleagues compared these normal control images with the ones taken of the players before the season started, they were astonished. Although none of the players in the control group had ever been diagnosed with a concussion, their scans showed long-lasting brain changes similar to the changes in those who had been previously diagnosed. Nauman at first thought the university's scanner was broken. Then he realized what he had stumbled on: there are more than one million high school football players who hit the gridiron every year, and upwards of sixty thousand of them will be diagnosed with a concussion. In actuality, there are likely to be at least twice that number who suffer concussions, who have all the symptoms, but who shake them off and either disregard the seriousness of the injuries or simply fail to report them to the coaches and trainers for fear of being taken out of the game. Some experts, in fact, believe there are closer to a quarter million concussions sustained by high school football players every year. When you add in tens of thousands more who have never been symptomatic but who have nonetheless suffered an injury resulting in brain damage, then the magnitude of the problem and the risk to our teenage athletes are staggering.

The damage to the brain is also not easy to detect because it is usually not structural. Rather, it is cellular, and yet it is severe enough to interrupt normal functioning and cause physical and cognitive symptoms, some of them immediate, others delayed for days, weeks, even months. Essentially, after the brain is violently moved inside the skull, a biochemical onslaught of calcium and potassium floods the brain, causing two things to happen. First, these chemicals, in excess, damage and destroy brain cells. Second, to pump out these excess chemicals, the brain needs its chief energy source, glucose. Normal blood flow in the brain carries the glucose to where it's needed, but after a concussion glucose distribution is restricted for several reasons. The calcium flood causes the brain's blood vessels to constrict and interferes with the breakdown of glucose, which is necessary for the production of energy, and as the brain swells with the influx of calcium and potassium, the blood vessels are further constricted. Not only are neurons affected, but white matter takes a "hit," too. The white matter tracts are stretched by the impact force, and can get "sheared." Experimental research using traumatic brain injury (TBI) models in rats and mice shows that the immature adolescent brain is very vulnerable to injury, and even mild injury can be associated with loss of synapses. In addition, there is a decrease in the NMDA type of glutamate receptor, which is required for LTP and memory. This may contribute to learning problems seen after concussion.

The physical side effects of concussion include dizziness, headache, blurry vision, sensitivity to light or sound, problems with balance, fatigue and lethargy, and a change in sleep patterns—either too much or too little sleep. The cognitive effects of concussion include amnesia, slow or fuzzy thinking, an inability to concentrate, and an inability to remember new information. A concussion can also result in mood changes, with the person becoming sad or irritable, nervous or anxious. When there are lingering symptoms—for weeks, months, even years—then a postconcussive syndrome is usually diagnosed.

The risk of a postconcussive syndrome and serious brain damage increases dramatically when an athlete suffers a second concussion before the symptoms of the first have fully resolved. All those energy-starved cells that are still trying to bounce back from the initial flood of calcium and potassium ions can be at risk of further injury. This is called second-impact syndrome.

A few years ago the *New York Times* told the story of Sarah Ingles, a high school basketball player in Ohio who sustained a concussion during a basketball game. Two hours later, while riding back to school on the bus with her teammates, she suddenly didn't know where she was and had no memory of even playing in the game. This was her second concussion, and it caused her to miss six weeks of school. After graduation, Ingles attended Ohio Wesleyan University, where she continued to play field hockey—until suffering her seventh concussion. (Not all of the five concussions in between were the result of sports. One came from bumping her head on a bed frame.) In a video she made for the Midwest's North Coast Athletic Conference while she was a student, she said that trainers and doctors told her she could no longer participate in contact sports, even at the intramural level. For months afterward she had repeated cognitive testing and was given a tutor to help her study. "Retaining information wasn't possible," she says in the video. "I had to go over it and over it and over it."

According to the most recent research, high school athletes who suffer three or more concussions are at an 8-fold increased risk for loss of consciousness after a successive concussion and a 5.5-fold increased risk for anterograde post-traumatic amnesia, in which the adolescent has trouble forming new memories. Recently, many neurologists and researchers have said it is more accurate to call a concussion a mild traumatic brain injury. There are approximately 1.5 million incidents of traumatic brain injury reported in the United

States every year. Seventy-five to 95 percent of these injuries are categorized as mild.

Second-impact syndrome can also be deadly. It was for Nathan Stiles, a seventeen-year-old star running back on his Spring Hill, Kansas, football team. In 2008, Nathan complained of a lingering headache after a game, and the team's athletic trainer advised his parents to take him to the emergency room. A brain scan at the hospital showed no problems, but just to be safe, doctors advised him to sit out the next three weeks. Without their permission, Nathan was legally prevented from playing since Kansas is one of dozens of states that require a physician to clear a high school athlete for play after the athlete has suffered a head injury. Three weeks later, the doctor did. In Nathan's first game back his mother watched him get hit and act a bit stunned. Afterward he said he was fine. Nothing seemed amiss the following week. It was the last football game of the year, the last of Nathan's senior season, and, as it would turn out, the last of his life. Right before halftime, Nathan, playing defense, intercepted a pass and ran it back for a touchdown. On the sideline he collapsed, then began suffering seizures. He was airlifted to the University of Kansas Medical Center and underwent four hours of brain surgery to stop the bleeding in his brain. He never woke up and was taken off life support the next day.

After an autopsy, Nathan's brain was sent to Boston University's Center for the Study of Traumatic Encephalopathy, where diseases and disorders resulting from brain injuries are cataloged and analyzed. Boston University's program works closely with the Veterans Affairs Center in nearby Bedford, Massachusetts, and when pathologists there opened Nathan Stiles's teenage brain, they were horrified at what they saw—a young brain filled with tau protein, the twisted fibers that choke and kill brain cells in Alzheimer's patients. There was only one explanation: At the time of his death, Nathan was suffering from chronic traumatic encephalopathy (CTE), a progressive degenerative disease normally found in athletes who have suffered repeated brain trauma. The disease was first identified in retired, elderly boxers in the 1920s. Nathan probably had suffered multiple undiagnosed concussions and may have had a genetic predisposition to concussions. To date, his is the youngest case of CTE ever recorded.

There is no concussion-proof helmet, and as doctors and researchers have recently realized, the damage in a young brain from a concussion can just as easily be caused by a subconcussive blow to the head. In 2011, the Canadian Paediatric Society declared that concussions were too common in youth sports, especially ice hockey, and that hits to the head, as well as fighting and checking from behind, should be banned in all youth sports. Currently bodychecking, where one player is allowed to slam another into the boards, is allowed even at the peewee level, ages eleven to twelve.

In 2012, Canadian researchers confirmed what many neuroscientists had been saying for some time, that a child's or teenager's brain is not as resilient as an adult's. Their subjects were ninety-six teenage athletes (rugby, hockey, football) between the ages of thirteen and sixteen who had suffered a concussion sometime in the previous six months. Using standard neuropsych tests, the scientists evaluated the teens' working memory abilities. Working memory is short-term memory, and it's important to the smooth functioning of the prefrontal cortex, which helps us when we read, remember a phone number, or do a simple mental calculation. Compared with a control group of similar-aged subjects who had no history of concussion in the prior six months, the recently concussed athletes had markedly worse short-term memories.

Radiologists at New York University also have found mood swings, sleep disorders, obsessive-compulsive behavior, anxiety, and impulse control disorders in patients with postconcussive syndrome following a mild traumatic brain injury. And other studies have found 15 to 20 percent rates of depression in patients up to a full year after a concussion; these are greater than the rate of depression in the general population.

Although it's not understood why, adolescent athletes take longer to recover from mild traumatic brain injury than adults do. The younger the athlete, the longer it takes. On average, adults need three to five days to return to baseline on cognitive tests; college athletes need five to seven days, and high school athletes ten days to two weeks. In one survey, more than half of high school players, average age sixteen, took more than a week to recover, and 10 percent took longer than three weeks. In subtler brain-imaging tests, brain abnormalities have been discovered in young athletes who were no longer symptomatic. One of those studies, involving late-adolescent athletes with a history of concussion, found brain abnormalities more than three years after their most recent injury.

Today, the NCAA, which oversees college athletics, maintains guidelines for how concussions should be handled during a game. For instance, schools do not allow student-athletes to just "shake it off" and return to the game without being evaluated by a health professional. But many people, including many former college athletes, believe the NCAA has not gone far enough. In 2011 four of those former student-athletes filed a class action suit against the NCAA, claiming the organization failed to implement appropriate concussion screening, return-to-play guidelines, and other safety measures. (The suit is still working its way through the judicial system.)

Recently, a study by the Behavioral Health Services at Nationwide Children's Hospital found that children and adolescents who suffer mild traumatic brain injury may be more likely to show an increase in symptoms, both cognitive and physiological, over time than children and adolescents who experience an orthopedic injury. Those cognitive and physiological symptoms were also associated with declines in physical and psychosocial quality of life.

Young people with severe brain injuries may be at risk for developing, during their recovery, what neurologists call a neurocognitive stall, which is a slowing of cognition as well as of social and motor development beyond a year after the injury. When they hit the wall or plateau, later developmental milestones are suddenly in jeopardy, especially in younger patients.

There is also a risk, although small, of developing epilepsy following a concussion just like Holly, the fifteen-year-old daughter of the woman who e-mailed me in 2010. Holly was a rare case in part because the sport she had been playing was boys' football. Among young adults, however, post-traumatic epilepsy is the most common cause of new epilepsy in the United States. The exact neurological events are still unclear. For years researchers believed it was simply overly stimulated neurons that created epileptic seizures, but more recently scientists have discovered evidence suggesting that before neurons even become overexcited, the damage is caused by an influx of neurochemicals trying to repair a brain injury. These chemicals, the researchers say, are what cause the excitation that results in the neuronal damage.

Concussions can cause physiological damage as well as cognitive damage, especially to the pituitary, a teardrop-shaped gland at the base of the hypothalamus. Because it is located behind the bridge of the nose, the pituitary is injured from even low-level impacts to the head. Known as the "king" gland, the pituitary is responsible for metabolism, growth, and energy, and some studies suggest that close to 40 percent of teenage athletes who suffer concussion also incur sport-related post-traumatic hypopituitarism, which includes a host of symptoms:

- Reduced muscle mass
- Weakness
- Decreased exercise capacity
- Fatigue
- Irritability
- Depression
- Impaired memory
- Reduced sex drive

Researchers are still trying to figure out the specifics of why adolescents appear to suffer longer-lasting consequences after concussion, but the fact that adolescent brains are still maturing must certainly play a large role. When teenage brains take a hit, the injury isn't static. Because the teenage brain is still developing, the injury is a trauma not just to a piece of gray matter but also to what would have been had the brain continued to develop without incident. Teenagers are damaging more than just their brains with concussions. They're damaging their futures.

I've heard many startling stories from my colleagues about young people and concussion, but the most heartbreaking ones come from adolescents whose multiple concussions have left them afraid for their future. Sarah Ingles, the Ohio Wesleyan University student-athlete who suffered seven concussions from high school through college, said in her video interview for her school's athletic conference, "Athletics was pretty much what I did." Her first warning sign came with her first concussion, when her head violently hit the turf in a high school field hockey game and she couldn't remember her name. Her second concussion, toward the end of the basketball season just a few months later, wasn't even a hard impact, she said, but she still felt confused on the bus ride home. Even before her seventh concussion, Sarah was plagued with headaches and nausea, but worst of all were the problems she had just trying to think clearly. When doctors told her she should never again play contact sports, Sarah took the advice and started playing golf: "If I got hit, I'd be in the hospital. . . . It's not only affected me in sports, but in life. . . . Head injuries are head injuries. It's your brain."

16
Crime and Punishment

Growing up in Florida, Terrance Graham was athletic and intelligent, but he had a troubled background. His parents were addicted to crack cocaine, and in grammar school he was diagnosed with attention deficit hyperactivity disorder. By the time he was nine years old, he was smoking cigarettes and drinking alcohol, and by age thirteen was using marijuana. In July 2003, when he was sixteen, Graham and three friends attempted to rob a barbecue restaurant where one of the friends worked. It was this accomplice who purposely left the back door unlocked at closing time, allowing Graham and one of the other boys to slip in. When they surprised the restaurant's manager, Graham's accomplice hit the man with a metal pipe, cutting his head. When the manager continued to yell for help, the three boys ran out the back of the restaurant, where a fourth youth was waiting in a car. No money was taken, the manager required stitches to his head, and within a couple of days all four were arrested. Prosecutors decided to charge the boys—none of them eighteen—as adults. Graham, who had never been arrested before, faced two felony counts. The most serious, a first-degree felony, carried a maximum penalty of life in prison without the possibility of parole.

Graham was offered a plea agreement and admitted his part in the crime. He was sentenced to a year in the county jail, less the six months already served. In June 2004 he was released on probation, intending, as he told the court at sentencing, to turn his life around. "This is my first and last time getting in trouble," he told the judge. His promise lasted six months. In December 2004, Graham, then seventeen, and two twenty-year-old accomplices forced their way into a man's home and held him hostage at gunpoint while they ransacked the house, looking for money. Later in the evening the three attempted another home invasion and robbery, during which one of Graham's friends was shot. Driving his father's car, Graham dropped the man off at the hospital and then sped past a policeman. A short time later he crashed into a telephone pole, tried to run, but was caught and arrested. It was December 13, 2004, and Graham was thirty-four days shy of his eighteenth birthday. In what was essentially a one-day trial in front of a judge, evidence was presented that Graham had violated his probation. He was facing a wide range of possible sentences under Florida law, running from a minimum of five years to life in prison. The Florida Department of Corrections recommended four years; prosecutors recommended thirty. At sentencing, Circuit Judge Lance M. Day of Duval County told Graham, "I don't know why it is that you threw your life away. . . . The only thing that I can rationalize is that you decided that this is how you were going to lead your life and that there is nothing that we can do for you. . . . I have reviewed the statute. I don't see where any further juvenile sanctions would be appropriate. . . . Given your escalating pattern of criminal conduct, it is apparent to the Court that you have decided that this is the way you are going to live your life and that the only thing I can do now is to try and protect the community from your actions."

Indeed, in 2004 the US Supreme Court debated the extent of accountability of adolescents. During the case, it was asked whether a juvenile was akin to a mentally retarded adult, without full faculties to know right from wrong.

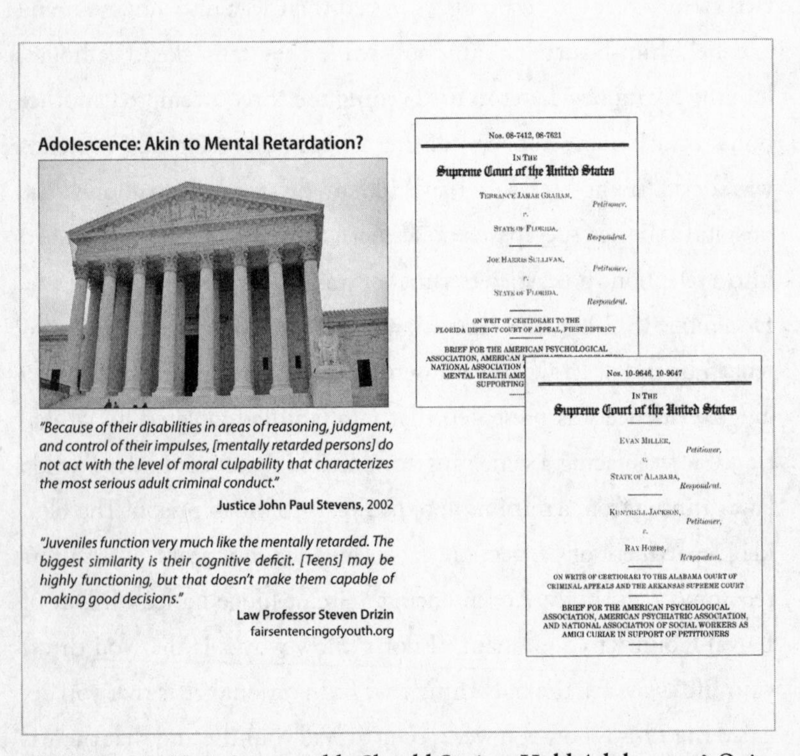

FIGURE 29: How Accountable Should Society Hold Adolescents? Opinions on the treatment and rehabilitation potential of juveniles who commit capital crimes vary widely.

With that, and on the basis of earlier charges of attempted armed burglary and attempted armed robbery, the judge sentenced Terrance Jamar Graham, age nineteen, to life in prison. Because Florida no longer has a parole system, a life sentence means exactly that—no possibility of release, unless by executive clemency. For a serious but nonhomicidal crime committed when he was a juvenile, Graham would be likely to spend the next sixty or seventy years in prison and die there. His attorneys appealed the sentence, citing the Eighth Amendment: "Excessive bail shall not be required, nor excessive fines imposed, nor cruel and unusual punishments inflicted."

In the spring of 2008 I was contacted by the Washington, DC, branch of the law firm of Clifford Chance. The firm was preparing an amicus brief for Graham's attorneys, who had appealed their client's life sentence all the way to the US Supreme Court. Now the firm needed experts to explain why adolescents should be held to a different standard. Of the seventeen who joined in the amicus brief, I was the only neurologist. Thus began my adventure in the intersecting worlds of juvenile justice and neuroscience.

The lawyers for Graham were arguing the same principle that led the Supreme Court to rule the death penalty for juveniles unconstitutional in 2005. Central to that decision was the idea that there is something unique about the intellectual, emotional, and psychological makeup of adolescents. The lawyers who put together the amicus brief summed it up in the first paragraph:

> Although adolescents must be held responsible for their actions, they generally lack mature decision-making capability, have an inflated appetite for risk, are prone to influence by peers, and do not accurately assess future consequences.

For the better part of this book I've provided scientific evidence supporting the notion that teen brains are different from adult brains. The question the high court was considering was how to weigh those differences in sentencing people for crimes committed when they were still adolescents.

Every year about 200,000 youths between the ages of twelve and seventeen are arrested for violent crimes. In 2008 juvenile offenders were involved in more than nine hundred murders and accounted for 48 percent of all arson in the nation. In 2005 the Supreme Court ruled it was unconstitutional to sentence to death anyone who was under the age of eighteen at the time the crime was committed. Seventeen years earlier it ruled the death penalty unconstitutional for anyone under the age of sixteen. But sentences of life without the possibility of parole (as well as the minimum age for criminal prosecution) have largely been left up to the states. Only three states do not allow charging youths under the age of seventeen in adult courts, and of the forty-seven that do allow it, twenty-nine actually mandate it for certain offenses.

Of all the countries in the world, the only two not to have signed the United Nations Convention on the Rights of the Child, a binding international treaty outlawing the sentencing of juveniles to life in prison without parole, are Somalia and the United States. It's a dubious distinction, to be sure. In America, of the approximately 40,000 prisoners currently serving sentences of life without parole, some 2,500 are offenders who were below the age of eighteen at the time they committed their crimes. Seventy of those prisoners were only thirteen or fourteen.

The overwhelming majority of the crimes for which juveniles are sentenced to life without parole are homicides, although nationally there are more than a hundred juveniles like Terrance Graham serving life without parole for nonmurders, including six who were thirteen or fourteen when they committed their nonhomicide crimes.

For most of human history, children have been regarded simply as pint-size adults—and more often than not treated as indentured servants. In the oldest known written legal document, the Code of Hammurabi, which dates to approximately 1780 BC, children were subject to the same rules as adults and yet at the same time were under the complete control of their fathers. If a son struck his father, according to the Code, the father could cut off the son's hands. Ancient Roman cultures continued the tradition of a father's total control over his children, and during the Middle Ages the state often acted in place of the parent. Children who were tried for crimes as adults were punished as adults. Documents from the Middle Ages make references to boys and girls as young as six years old being hanged or burned at the stake. By the middle of the sixteenth century, however, more thought was given to the idea that criminals, and especially criminal children, could be taught to change their delinquent ways. In April 1553, Bridewell, an old manor house of the British monarchy, was given over to the city of London by King Edward VI at the urging of Bishop Ridley, who proclaimed in a sermon the need for a workhouse for the poor and a house of correction "for the strumpet and idle person, for the rioter that consumeth all, and for the vagabond that will abide in no place." Bridewell Prison subsequently became a human warehouse not only for the indigent, homeless, and lame but also for vast numbers of children accused of committing petty offenses.

The first record of a juvenile being put to death in America dates to the seventh of September 1642 and poor sixteen-year-old Thomas

Granger, who was found guilty of the high moral crime of sodomy with farm animals. According to William Bradford, the governor of the Plymouth colony, Granger, "late servant to Love Brewster of Duxborrow, was [in] this Court indicted for buggery wth a mare, a cowe, two goats, divers sheepe, two calves, and a turkey, and was found guilty, and received sentence of death by hanging untill he was dead."

The Church of Rome was one of the first institutions to set a legal demarcation between adults and juveniles when it proclaimed children under the age of seven unable to form intent in the commission of a crime. Pope Clement XI followed through on his proclamation in 1704 when he founded a center for "profligate youth" in Rome—perhaps the Western world's first reformatory school. But it took England's most acclaimed legal scholar, William Blackstone, to lay the groundwork for juvenile justice systems in both Great Britain and America. After giving a series of lectures about the law at Oxford University in 1753, Blackstone codified the principles he had espoused into a four-volume legal work called *Commentaries on the Laws of England*, which was published between 1765 and 1769. In the *Commentaries* he describes the two criteria necessary to hold someone accountable for a crime. The first was obvious: A person had to have committed an illegal act. The second was much more difficult to ascertain: A person had to have a "vicious will," by which Blackstone meant the intent to commit a crime. He then goes on to describe different categories of people in whom a "will," by definition, is lacking. The first group he singles out is "infants," and by "infants," Blackstone simply meant children—that is, children below the age at which they could fully understand the meaning and consequences of their actions. And exactly what age was that? Actually there were two: children ages six and younger, as a rule, could not be found guilty of a serious crime, while children fifteen and older could be tried, convicted, and sentenced as adults. And if you were ages seven to fourteen, then what? Blackstone fudged. He said if an "infant" could understand the difference between right and wrong, then the child should be held accountable in the same manner as an adult.

> The capacity of doing ill, or contracting guilt, is not so much measured by years and days, as by the strength of the delinquent's understanding and judgment. For one lad of eleven years old may have as much cunning as another of fourteen; and in these cases our maxim is, that "malitia supplet aetatem" ["malice supplies the age"].

That the courts, no less than the public, remained confused about the age at which children should be deemed capable of understanding right and wrong and the consequences of their actions is evident in the fact that the executions of children younger than fourteen continued in the United States of America well into the eighteenth century. In New London, Connecticut, on December 20, 1786, Hannah Ocuish, a twelve-year-old girl, part Pequot Indian, part African American, was hanged for beating and strangling to death a six-year-old white girl from a wealthy family because, according to Hannah's tearful confession, the six-year-old "complained of her . . . for taking away her strawberries." Before Hannah's execution, local Unitarian minister Henry Channing, who counseled the young girl during her imprisonment, preached to the throng that had assembled to witness Hannah's hanging. In a sermon titled "God, Admonishing His People of Their Duty as Parents and Masters," Channing spoke of the "natural consequences of too great parental indulgence" and warned parents that "appetites and passions unrestrained in childhood become furious in youth; and ensure dishonour, disease and an untimely death." As Hannah awaited death, Channing thundered, "Sparing you on account of your age, would, as the law says, be of dangerous consequence to the publick, by holding up an idea, that children might commit such atrocious crimes with impunity." It didn't help Hannah that she was part Indian and part African American and that she had killed a white child. Shortly after noon, Hannah was hanged from the gallows by the sheriff. So popular was Channing's sermon that within days he was named pastor of New London's First Church of Christ.

Some, however, were horrified by Hannah's execution, not only because of her age—she is considered to be the youngest female ever executed in the United States—but also because she was thought by some at the time to be mentally handicapped. She was also the product of a broken home and an alcoholic mother who had given her away to become a servant to a wealthy white family. It's not clear how well Hannah understood the sentence she received after her one-day trial, but when she confessed to the crime—after being taken to view the six-year-old's dead body—she cried and swore she was sorry and would never do it again.

The push to find some alternative means of dealing with children who commit criminal acts reached a turning point in 1825 when New York City's "House of Refuge" for children was established. It was the first official institution in the United States dedicated solely to protecting and rehabilitating juvenile offenders and therefore the first tacit recognition by society that there might be special circumstances contributing to some juvenile crimes. Prior to that time in New York, youthful criminals had been housed in the state penitentiary, which opened in 1797. The House of Refuge took charge not only of juvenile criminals but also of orphans, poor children, and any youth considered to be "wayward."

> Any child who for any reason is destitute or homeless or abandoned; or dependent on the public for support; or has not proper parental care or guardianship; or who habitually begs or receives alms; or who is living in any house of ill fame or with any vicious or disreputable person; or whose home, by reason of neglect, cruelty or depravity on the part of its parents, guardian or other person in whose care it may be, is an unfit place for such a child; and any child under the age of eight who is found peddling or selling any article or singing or playing a musical instrument upon the street or giving any public entertainment.

The children of the House of Refuge spent eight hours a day working and learning trades such as tailoring and brass nail manufacturing, with another four hours given to more traditional schooling. Within fifteen years there were dozens of Houses of Refuge across the United States; these houses eventually spawned reform and training schools, beginning in 1886 with the Lyman School for Boys in Westborough, Massachusetts. The first true juvenile court, however, wasn't established until 1899, in Cook County, Illinois, after the state legislature passed the Juvenile Court Act. One of its first judges, Julian Mack, described how a separate justice system for juveniles was different from adult criminal courts:

> The ordinary trappings of the courtroom are out of place in such hearings. The judge on a bench, looking down upon the boy stand-

> ing at the bar, can never evoke a proper sympathetic spirit. Seated at a desk, with the child at his side, where he can on occasion put his arm around his shoulder and draw the lad to him, the judge, while losing none of his judicial dignity, will gain immensely in the effectiveness of his work.

Responsible for youths charged with major crimes, juvenile courts also meted out punishment for such things as vagrancy and truancy. And because these were special courts, without juries or even due process, children could be sent away simply on the orders of a single judge and on the basis of evidence that did not need to be beyond a reasonable doubt. In the early part of the twentieth century children could be taken into custody just for throwing snowballs and summarily sent away to a detention facility.

Overcrowding and abuse in reform and training schools brought juvenile "rehabilitation" into the discussions of civil rights activists in the 1960s and ultimately, in two landmark cases, to the US Supreme Court. The first case, in 1966, was *Kent v. United States*. Morris Kent entered the juvenile justice system in Washington, DC, at the age of fourteen for minor offenses, including attempted purse snatching. But when he was sixteen, his fingerprints were found in the apartment of a woman who had been robbed and raped. It was in the juvenile court judge's discretion to send the case to adult criminal court, and he did so without a hearing, refusing to hear arguments from Kent's lawyers. When the high court heard the case, it ruled that Kent had been denied "meaningful representation" and due process. Justice Abe Fortas wrote in the majority opinion "that there may be grounds for concern that the child receives the worst of both worlds:

that he gets neither the protections accorded to adults nor the solicitous care and regenerative treatment postulated for children."

By the time the US Supreme Court made its ruling, Kent was twenty-one and could not be retried in juvenile court. Instead, his case was remanded to DC's district court with instructions to consider the Supreme Court's ruling and assign punishment "consistent with the purposes of the Juvenile Court Act." Kent was found guilty of six counts of housebreaking and robbery, but not guilty by reason of insanity on the two counts of rape. He was treated at Saint Elizabeth's, a psychiatric facility in DC, from 1963 to 1968, at which time he was declared sane. Eventually Kent married and had children, and apparently he has lived a crime-free life since.

The same year as the Kent ruling, 1966, Fortas wrote the opinion for another landmark case in juvenile justice, *In re Gault*, which involved Arizona fifteen-year-old Gerald Gault. On June 8, 1965, the sheriff of Gila County took Gerald and a friend into custody. At the time, the teenager was on probation for being in the company of a boy who had stolen a woman's wallet from her purse. When the sheriff grabbed him, Gerald was accused of making lewd phone calls, which Fortas described this way a year later in the court's majority opinion: "The remarks or questions put to her were of the irritatingly offensive, adolescent, sex variety." After being arrested, Gerald was taken without a hearing to the Children's Detention Home, where he remained for several days. When he was released, the sheriff presented Gerald's mother with a written note saying the judge would be hearing Gerald's case the following week, at which time the young boy, who was not represented by a lawyer, was sentenced to a juvenile correctional facility for an undetermined time not to exceed his twenty-first birthday. Under Arizona law, juvenile sentences could not be appealed. The following year, the Supreme Court ruled that Gault's arrest, detention, and sentencing were all in violation of the US Constitution. Fortas wrote, "Under our Constitution, the condition of being a boy does not justify a kangaroo court." Gault was eventually released and worked at a series of odd jobs for a few years. He served in the US Army from 1969 to 1991 and several years ago was working toward his teaching credentials.

Adolescence, at this time, was still widely understood as a psychological and physiological stage of life, not as a "brain stage." A discussion about the relationship between biology and legal accountability was forced on the criminal justice system following a horrifying mass murder on August 1, 1966. Twenty-five-year-old Charles Whitman was a former altar boy, an Eagle Scout, a US Marine, and a college student when he killed his wife and his mother in their beds and then, from the observation deck of the twenty-eight-story University of Texas Tower, began indiscriminately firing a high-powered rifle at the people below. Among those he killed on that sultry August afternoon were a pregnant woman, an eighteen-year-old student and his fiancée, an electrician, a professor, and an ambulance driver who had arrived to help the wounded. Before he was shot and killed by police, Whitman murdered thirteen and wounded thirty-two. Among his four suicide notes, written over a period of hours, the one dated 6:45 p.m., Sunday, July 31, 1966, the day before the massacre, read:

> I don't quite understand what it is that compels me to type this letter. Perhaps it is to leave some vague reason for the actions I have recently performed. I don't really understand myself these days. I am supposed to be an average reasonable and intelligent young

> man. However, lately (I can't recall when it started) I have been a victim of many unusual and irrational thoughts.

It is noteworthy that Whitman referred to himself as a "victim" because he obviously felt he was not in control of his violent urges. Shortly before his death he visited the university doctor about his depression and headaches and was prescribed Valium. The doctor suggested Whitman also make an appointment with the university psychiatrist, which he did. Whitman, however, was convinced that something must be physically wrong with him, so he made this request in his suicide note:

> After my death I wish that an autopsy would be performed on me to see if there is any visible physical disorder. I have had some tremendous headaches in the past and have consumed two large bottles of Excedrin in the past three months.

An autopsy was performed, and an explanation was found—a grade 4 glioblastoma multiforme tumor about the size of a walnut protruded from under the thalamus and impinged on both the hypothalamus and the amygdala. Was Whitman unable to harness his aggression or inhibit his destructive impulses because of the tumor's effects on his brain? While it's not possible to say for certain, the tumor certainly could have interfered with the normal balance of activity between his frontal lobes and limbic system. Limbic overdrive can result in rage episodes, and if the connections that normally inhibit or dampen these impulses were damaged by the presence of the tumor, he may have been incited to commit these crimes and lacked the ability to control those dangerous impulses. This is an extreme example where brain connectivity may have played a role, but the idea that reduced frontal lobe connectivity, now seen as part of the maturing process in adolescents, could or should mitigate our understanding of juvenile justice was still years away. The case of *Roper v. Simmons* helped.

Christopher Simmons grew up in Jefferson County, Missouri, where he suffered continual physical and emotional abuse at the hands of his stepfather. When Simmons was only four years old, his stepfather took him to a bar and fed him alcohol to entertain the other patrons. He was once hit so hard in the head by this man that his eardrum burst. By the time he reached his teens, Simmons was regularly drinking alcohol and smoking marijuana and occasionally indulging in harder drugs. To escape the abuse, the teenager hung out at the trailer home of a neighbor, a twenty-eight-year-old man who gave him more drugs and encouraged him to steal and split the profits.

On September 9, 1993, Simmons, then seventeen, and a younger friend decided to commit a heinous crime. Picking Shirley Crook's home at random, they bound and gagged the forty-six-year-old woman, robbed her, then drove to a state park and threw her off a bridge into the Meramec River, where she drowned. Simmons was arrested; he confessed and was tried and convicted of first-degree murder, and then sentenced to death. The constitutionality of the sentence was questioned on appeal, and eventually the case reached the US Supreme Court. On March 1, 2005, the high court called the execution of a minor a violation of the Eighth and Fourteenth Amendments of the US Constitution, as well as "the evolving standards of decency that mark the progress of a maturing society." Writing for the majority, Justice Anthony Kennedy stated:

> When a juvenile offender commits a heinous crime, the State can exact forfeiture of some of the most basic liberties, but the State cannot extinguish his life and his potential to attain a mature understanding of his own humanity.

Fourteen months after that historic decision, in Jacksonville, Florida, Terrance Graham was sentenced to life in prison without the possibility of parole for a home invasion robbery. And two years after that, I received the phone call from the law firm representing Graham. As the only neurologist contributing to the amicus brief, I felt an enormous sense of responsibility to convey all that I knew about the limitations of the teenage brain when it comes to controlling impulses, assessing risk, avoiding peer pressure, and understanding the consequences of one's actions. Beyond that, it was critical to describe how adults and adolescents use different parts of their brains when acting and reacting in certain situations.

One distinguished legal scholar, Steven Drizin of Northwestern University in Chicago, has even said, "Juveniles function very much like the mentally retarded. The biggest similarity is their cognitive deficit. [Teens] may be highly functioning, but that doesn't make them capable of making good decisions."

Teens, we now know, engage the hippocampus and right amygdala when faced with a threat or a dangerous situation—this is why they are prone to being emotional and impulsive—whereas adults engage the prefrontal cortex, which allows them to more reasonably assess the threat. We know that the risk factors for teens committing violent acts include seeing violence and being the victims of it themselves. We know that teens are much more likely to be influenced by peers when they take risks or engage in dangerous and/or criminal

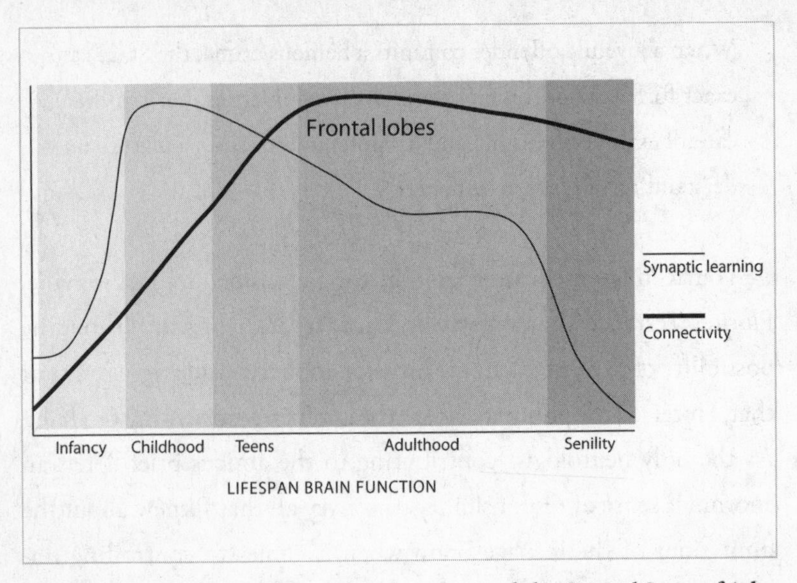

FIGURE 30. A Recap of Brain Development and the Critical Stage of Adolescence: Synaptic learning rises rapidly in infancy and childhood, stays high during the teen years, and tapers off to plateau in adulthood. However, myelination and connectivity do not peak until early adulthood (late twenties), plateauing well into later life. Learning abilities and rates tend to be highest in children and teens, decreasing along with synaptic function in late adulthood, although adolescents are challenged by their relative lack of connectivity of the frontal lobes, the last parts of the brain to connect. The teen years provide an exceptional opportunity to work on strengths and weaknesses.

behavior. (On average, half of all homicides committed by juveniles involve multiple accomplices.) We know that because of their still-maturing frontal lobes, adolescents have trouble understanding the consequences of their decisions, and therefore they are impaired in their ability to assess things such as their Miranda rights, the competency of their representation, and the consequences of plea bargaining. And finally, we now know that adolescents, especially those between the ages of twelve and seventeen, are particularly vulnerable to making false confessions.

In a 2004 analysis of 133 false confessions, researchers found that 16 percent came from sixteen- and seventeen-year-olds, the highest concentration of any age group. Valerie Reyna, a teacher and researcher in the Department of Human Development at Cornell University, summed up the competence of adolescents in the juvenile justice system when she wrote in a 2006 journal article:

> In the heat of passion, in the presence of peers, on the spur of the moment, in unfamiliar situations, when trading off risks and benefits favors bad long-term outcomes, and when behavioral inhibition is required for good outcomes, adolescents are likely to reason more poorly than adults do.

In our team's thirty-six-page amicus brief for *Graham v. Florida*, we wrote:

> Well established, growing, and uniform scientific and academic study shows that the purposes of a sentence of life without parole—punishing the culpable, deterring the sensible, and incapacitating the incorrigible—are not reliably or rationally served by the imposition of that sentence upon adolescents.

Thankfully, on May 17, 2010, the United States Supreme Court agreed. Justice Anthony Kennedy wrote, for the majority, "With respect to life without parole for juvenile nonhomicide offenders, none of the goals of penal sanctions that have been recognized as legitimate—retribution, deterrence, incapacitation, and rehabilitation . . . provides an adequate justification. . . . Even if the punishment has some connection to a valid penological goal, it must be shown that the punishment is not grossly disproportionate in light of the justification offered. Here, in light of juvenile nonhomicide offenders' diminished moral responsibility, any limited deterrent effect provided by life without parole is not enough to justify the sentence."

Although siding with the majority, Chief Justice Roberts took issue not with the new science behind our understanding of the still-maturing adolescent brain but rather with the Court's venturing an opinion on that science. "Perhaps science and society should show greater mercy to young killers," Roberts wrote, "giving them a greater chance to reform themselves at the risk that they will kill again. But that is not our decision to make."

The Graham case, however, did not fully protect adolescents from excessively harsh treatment by the criminal justice system. It left one thing unaddressed: sentences of life without parole in cases where a juvenile was convicted of homicide. Less than a year following the Supreme Court's majority decision in *Graham v. Florida*, the situation presented itself in the form of another case that had wound its way through the justice system and finally landed in the lap of the US Supreme Court. Our group, which had collaborated on the *Graham* amicus brief, was asked again to step in by the Clifford Chance law firm to examine the evidence and consider how neurobiology might be used to enlighten the justices of the Supreme Court. The firm sent out a kind of electronic call to arms in mid-2011 for us to once again provide social and biological science to argue that even in cases of homicide, adolescents under the age of eighteen should not be held responsible to the same degree as adults. The prevailing federal law was that, like adults convicted of murder, adolescents could be sentenced to life in prison without the possibility of parole. Our job was to point out that at least in some instances the juvenile may have acted on impulse owing to frontal lobe immaturity (nature) or as a result of a poor educational or stressful childhood environment (nurture). In other words, juveniles are less capable, neurologically, than adults of making mature decisions and are more vulnerable than adults to external influences. A corollary to our argument that there are clear neurological differences between adults and adolescents was the fact that teenagers' enhanced plasticity helps increase the odds in favor of their successful rehabilitation. In the amicus brief we wrote that "juveniles do typically outgrow their antisocial behavior as the impetuousness and recklessness of youth subside in adulthood. Adolescent criminal conduct frequently results from experimentation with risky behavior and not from deep-seated moral deficiency reflective of 'bad' character."

Up for argument were two cases involving juveniles sentenced to life without the possibility of parole for homicides committed when both boys were only fourteen years old. In one, *Jackson v. Hobbs*, Kuntrell Jackson from Arkansas tried to rob a video store in 1999 along with two older youths. One of the older youths shot and killed a clerk. In the second case, *Miller v. Alabama*, Evan Miller and a friend beat a fifty-two-year-old neighbor with a baseball bat and then set fire to his trailer. The middle-aged man died from blunt force trauma and smoke inhalation.

On June 25, 2012, at 3:22 in the afternoon, I received an e-mail from the Clifford Chance firm, thanking the amicus brief team:

> The Supreme Court handed down its decision today that it is unconstitutional to mandate a life sentence without the possibility of parole for all juveniles convicted of homicide under 18 years of age at the time of their crime. In doing so, the Court struck down

> statutes in 29 states with mandatory life-without-parole sentences that precluded consideration of the offender's age. . . . Congratulations on a job well done as your views were certainly heard!

All statutes that essentially required a person who committed murder as a child to die in prison were thus struck down. The ruling meant Kuntrell Jackson and Evan Miller, both of whom were given life in prison without parole for homicides they committed at the age of fourteen, would receive new sentencing hearings. Writing for the majority in the split 5-4 decision, Justice Elena Kagan said the problem with mandatory sentences was the inability to distinguish between, say, a seventeen-year-old and a fourteen-year-old, "the shooter and the accomplice, the child from a stable household and the child from a chaotic and abusive one." Kagan went on to cite the neurological evidence that had helped her to make her decision:

> Mandatory life without parole for a juvenile precludes consideration of his chronological age and its hallmark features—among them, immaturity, impetuosity, and failure to appreciate risks and consequences.

Still, the Court was once again deeply divided, pointing to the complexity of applying neuroscience to legal culpability. In fact, the justices also did not categorically ban juvenile life sentences without parole but rather, as Kagan wrote in the majority opinion, determined that "given all that we have said in *Roper*, *Graham*, and this decision about children's diminished culpability, and heightened capacity for change, we think appropriate occasions for sentencing juveniles to this harshest possible penalty will be uncommon." The

Court ruled that a judge or jury must have the opportunity to consider mitigating circumstances, *including the offender's age at the time of the crime*, before imposing the harshest penalty for juveniles.

Although each new ruling by the Supreme Court has reaffirmed that juveniles are constitutionally different from adults and must therefore be punished differently, the Court has left it up to the states to decide whether its rulings should be applied retroactively for all those inmates now in prison who were sentenced to life terms without parole for crimes committed when they were adolescents. (Currently, Michigan, Iowa, and Mississippi have agreed to apply the ruling retroactively.)

The law is finally taking into account the immature brains of adolescent criminals when it comes to sentencing, but as a society we're still a long way away from coming to grips with how to deal with juvenile criminals who suffer from mental disorders. (We certainly haven't figured this out for adult criminals either.) Part of the problem is distinguishing between an immature brain and a disordered brain, and as Chapter 12 described, this can sometimes be difficult. There is no disputing that news of horrific juvenile crimes, including mass shootings, seem all too common today. But the reach of the mass media and their saturation of our daily lives obscures the fact that juvenile violent crime rates, which did increase between 1985 and 1994, actually decreased by half between 1995 and 2011. The American Academy of Psychiatry and the Law cites risk factors for adolescent violence that range from access to firearms and drugs to poverty, parental neglect, family conflict, and exposure to violence on television, in movies, and through the Internet. What is also clear, however, is that the prevalence of mental illness is greater among adolescent criminals than in the general population. At the same time, the majority of juveniles who commit crimes are *not* mentally ill. The more troubling statistic in this regard may be that of those adolescents who *are* mentally ill, less than 25 percent get treatment.

The central problem is this: Despite the great advances in neuroscience, what we still don't know about the human brain dwarfs what we do know. Making judgments, even scientific judgments, based on what is available and known is at best foolhardy and at worst dangerous. That is certainly the case when it comes to pointing to objective evidence for a causal relationship between neuromaturity and real-world activity, especially criminal behavior. Brain scans have the appearance of hard-and-fast empirical data, but brain scans must be interpreted and are therefore far from objective. Scientists studying brain images, especially fMRI images, must weigh the technique used, the clarity of the image, and the choice of subjects in the sample or survey, and then finally must resist the temptation to find a one-to-one correspondence between a given brain region and a particular cognitive function. There is simply no such correspondence. Every brain region we tap—in fact every decision we make and every mood we feel—involves multiple cognitive processes. "Further hindering extrapolation from the laboratory to the real world is the fact that it is virtually impossible to parse the role of the brain from other biological systems and contexts that shape human behavior," wrote Jay Giedd, coauthor of an important article on adolescent maturity and the brain published several years ago. "Behavior in adolescence, and across the lifespan, is a function of multiple interactive influences including experience, parenting, socioeconomic status, individual agency and self-efficacy, nutrition, culture, psychological well-being, the physical and built environments, and social relationships and interactions." Giedd is chief of the Section on Brain Imaging in the

Child Psychiatry Branch of the National Institute of Mental Health. And while his list of influences on adolescent behavior is exhausting, it is not exhaustive. That's due, at least in part, to something I said earlier: that we are still learning when it comes to the human brain. Understanding the limitations of extrapolating from brain science to behavior is further complicated by the fact that increasingly scientists are stretching the age of neuromaturity. In other words, there is no bright line, no numerical age or boundary or demarcation at which we can say someone is neurologically mature. Instead, it is becoming increasingly clear that brain maturation extends well into a person's twenties. As a scientist and a physician I want to believe there are answers to every question and clear boundaries to every event and stage in life, but I know there are not. It might be tempting to think that once the tumult of the teenage years is over, it's smooth sailing, but that's hardly the case either. Meanwhile, our local governments continue to throw more money into building additional prison space and facilities rather than creating rehabilitation and counseling programs for adolescents at risk.

17
Beyond Adolescence
It's Not Over Yet

At the end of college most kids are still evolving. I witness this growth every day in my lab, where we take in college grads who are still undecided about whether they want to go to graduate school or medical school or whether they'd rather not pursue any postundergraduate degree at all and just get a full-time job out in the "real" world. They come in as bright, young students with excellent academic records to work as lab assistants. Their job is to help the other doctors and scientists with experiments and also keep the laboratory going with respect to cleaning and maintaining the equipment, and making sure orders are filled and meetings are scheduled. On entry, they embark on a steep learning curve concerning the science behind our work. However, the greater transformation they undergo comes subtly, as a result of having to communicate between lots of different types of people, organizing and prioritizing tasks, and knowing other people are counting on them. Over the course of two years or so, these young people really become adults; their sense of accountability and their communication skills in a professional environment are probably the greatest benefits they get from the whole experience. My own son Andrew did this very thing, leaving a physics master's degree program to change fields and work in a lab in New York City not much different from mine. From the other side, I saw his work ethic grow and his ability to organize and multitask get better, and like the kids in my lab, he got a huge boost in confidence from succeeding in a "real world" job. It will serve him well as he now enters advanced degree work.

In my lab, this growth shines through when these young people are about to leave, especially in their interviews for graduate school, medical school, or a job in the workplace. It's a joy for me to write their reference letters, and I marvel at the fact that this transformation happens, to at least some extent, to nearly every young person who spends a year or two in my lab. On occasion, some have brought their parents, who might be visiting from out of town, to the lab, and it gives me the greatest pleasure to tell them how proud they should be of their sons and daughters. Of course, I do manage to embarrass my young workforce this way!

What's important to remember is that young adulthood is still a great time to learn. There remains a high amount of brain plasticity going on, while brain connectivity has improved and so has the ability to multitask. Many young adults find that their learning skills are much better at this age than when they were in high school. Organizational skills improve, as does the ability to abstract. Judgment, insight, and perspective all improve as a result of more accessible frontal lobes.

The fact that many kids take a gap year prior to college makes a lot of sense developmentally. In many countries the "gap year" concept is nationally mandated. A number of European countries actually have a national "year of service" for teenagers when high school ends. In Israel, both boys and girls are drafted into the army. The boys serve three years, the girls two, and although those from certain Orthodox Jewish groups are exempt because of religious restrictions, they often engage in some other form of national service. The gap year for young Israelis comes the year *after* military service is finished and before they go on to a university. During that year many young people travel to Asia, South America, and India before they begin their higher education.

We, too, now have an informal "gap year" in the United States with community service or volunteer internships, such as the AmeriCorps program City Year, and testimonials from young adults describe how their gap year was the most valuable year of their lives. I witnessed this myself with my son Will, who took a year off after graduating from high school to travel solo around the world and to work. During that time he spent the first six months learning to speak Spanish in an immersion course in South America and then spent the remaining six months as an employee of a Boston software firm. By the time he was ready to enter Harvard, he was more mature both emotionally and intellectually, and he has told me more than once that the experiences of his gap year were invaluable to him in his development.

Young adulthood as a distinct developmental stage continues to be debated by sociologists, psychologists, and scientists. In the middle of the twentieth century the psychologist Erik Erikson was one of the first to suggest that adulthood had stages, broken down roughly into the ages twenty to forty-five, forty-five to sixty-five, and sixty-five to death. But it was Kenneth Keniston who, in 1970, wrote a seminal yet rather forgotten research paper called "Youth: A 'New' Stage of

Life." In it, Keniston postulated that a person's twenties—that is, the years between adolescence and young adulthood—were a distinct stage of development. Keniston was then a Yale psychologist, and he characterized this period chiefly as one of freedom, movement, change, and ambivalence. He also described a number of themes or issues that dominate this "youth" stage:

- Tension between self and society
- Pervasive ambivalence
- Wary probe
- Estrangement
- Omnipotentiality
- Refusal of socialization
- Youth-specific identities
- Movement
- Abhorrence of stasis
- Be moved/move through
- Valuation of development
- The fear of death
- View of adulthood
- Youthful countercultures

Keniston, who was writing about today's aging baby boomers, said that here was "a growing minority of postadolescents [who] have not settled the questions whose answers once defined adulthood: questions of relationship to the existing society, questions of vocation, questions of social role and lifestyle." The overriding characteristic of this "youth" stage, he said, was a "pervasive ambivalence toward self and society." This is a good news–bad news story. The energy and novelty-seeking so characteristic of this developmental stage are highly adaptive and allow teenagers to explore new domains that could match their skill sets, yet this same behavior can expose them to risky environments. Their lack of life experience can fuel ambivalence and fear, but this is where the family and community can come together around adolescents to provide reassurance and structure as they build their own context for their life experience.

As it turns out, it was an accurate description of my generation, but Keniston's theory of a postadolescent, preadult stage of life never really took hold—until psychologist Jeffrey Jensen Arnett wrote a book a decade ago called *Emerging Adulthood: The Winding Road from the Late Teens Through the Twenties.* Drawing on Keniston's work, Arnett postulated a distinct stage he called "emerging adulthood," mostly brought on by cultural and economic changes that have left twenty-somethings feeling insecure, needing more education, finding fewer jobs, resisting the rush to marry because of the acceptance of premarital sex, and delaying having children because of the sophistication of reproductive technology. The age of emerging adulthood is an age of exploration and instability, Arnett says, but today it's also an age of self-involvement.

That's certainly clear in the HBO series *Girls*, which features women in their twenties trying to figure out how to live, love, and succeed in New York City. It's worlds away from *Sex and the City*, and though it's billed as a comedy, there are plenty of uncomfortable and dramatic moments. The series was created in 2012 by then-twenty-six-year-old writer Lena Dunham, who also stars as the main character, an aspiring writer named Hannah. Early in the series, Hannah's parents tell her they can no longer subsidize her apartment in Brooklyn. When she then asks her boss to be paid, she is essentially fired from her unpaid internship. Later, she fails as a clerk in a law office and finally must settle for a job in a coffee shop. Hannah's life always seems to be in turmoil. She breaks up and makes up with her boyfriend multiple times and temporarily moves out of the apartment she shares with a female friend after they have an emotional disagreement. In other words, she's a mess—psychologically, emotionally, romantically, and vocationally. She's also not alone—hence the popularity of the series, especially among young women. In one episode, when she visits her upper-middle-class parents back in Michigan, Hannah catches up with a former friend, Heather, who tells her about her big dreams. Hannah is skeptical, but when she tells another friend about the conversation, it almost seems that she is talking about herself:

> Heather is moving to California to become a professional dancer, so that should make us all feel pretty sad and weird. . . . And nobody is telling her. She's going to go to LA and live in some shitty apartment, and she's going to feel scared and sad and lonely and weird all the time.

Hannah has moments of self-awareness, but they are few and far between. Instead, she and her friends vacillate about jobs, about men, about decisions both big and small in scenes that could just as easily be playing out among the approximately seventy million Americans who find themselves in this peculiar age, between eighteen and thirty-four, grown up and yet not quite grown up. The twenties are an age of self-absorption, of excitement and anxiety about the possibilities ahead, and especially of uncertainty—about jobs, careers, relationships; about who they are, where they're going, and when they'll get there. All across the developed world, young adults are trying to answer these questions. As they do, they are taking more time than ever before, at least in recent human history, to leave home, finish school, get married, and find a career. In some sense how do these kids *not* avoid being overly self-involved?

It's true that for the millennial generation, those born between 1980 and 2000, the incidence of narcissistic personality disorder is three times what it is for those sixty-five and older, according to the National Institutes of Health. Not surprisingly, the *Oxford English Dictionary* designated "selfie," those ubiquitous cell phone photos so many teens and young adults take of themselves, as its word of the year for 2013. In large part, many believe that the millennial generation's apparent self-centeredness was created by its parents, who, by way of overpraising their children, pushed them to become more self-involved and self-important. By instilling "too much" self-confidence in their kids, says Roy Baumeister, who teaches psychology at Florida State University, these parents "accidentally" injected their kids with a heavy dose of narcissism and a sense of entitlement as well. So there is a fine balance to strike when it comes to providing an assuring context to teens' experiences: too little and they feel lost, but too much and they are unrealistically confident, a setup for later problems in life.

On the positive side of the millennials' ledger, there is much to praise. While they are not as idealistic as past generations, they are earnest, pragmatic, and even optimistic despite growing up in the shadow of 9/11, two wars, and a debilitating recession. In her book *20 Something Manifesto*, Christine Hassler collects the reflections of postadolescent young adults who find this particular stage of their lives uniquely stressful. One twenty-five-year-old woman said it is "somewhat terrifying to think about all the things I'm supposed to be doing in order to 'get somewhere' successful: 'Follow your passions, live your dreams, take risks, network with the right people, find mentors, be financially responsible, volunteer, work, think about or go to grad school, fall in love and maintain personal well-being, mental health and nutrition.' When is there time to just be and enjoy?" Another twenty-five-year-old wrote, "Our culture really focuses on youth and success, and many of us feel that we have to be fabulously successful by age thirty or we're failures."

Laura Humphrey, a psychologist and director of Yellowbrick, a private psychiatric facility specializing in problems of emerging adulthood, describes the unique issues associated with the age group:

> The developmental agenda for all emerging adults is to define themselves, and their life's greater purpose, in relation to the larger community and world. They must do this as they redefine themselves within their family as increasingly independent while still emotionally connected. There is no greater developmental challenge in all of adult life.

Statistics seem to bear out the underlying insecurity of young adulthood: A third of all Americans in their twenties move every year, and 40 percent move back in with their parents at least once after college. Young adults go through an average of seven changes in employment before turning thirty, and two-thirds live with a partner without getting married for at least a portion of their twenties. Arnett claims that 60 percent of the subjects he's studied who are in their twenties say they feel both grown up and not quite grown up at the same time.

When your sons or daughters who are out of college and on their own seem unable to learn how to do laundry, create a budget, or set up utilities in a new apartment, remember, again, that while they're no longer adolescents, white matter is still being laid down in their frontal lobes, wiring their brain systems together. Like adolescents, young adults are sometimes victims of their own still-changing brains. That white matter connectivity still taking place after adolescence also carries substantial dangers. White matter abnormalities in the frontal regions of the brain appear to play a part in psychiatric disorders that develop not just during adolescence but also in young adulthood. In fact, mental disorders are estimated to account for nearly half the total disease burden for young adults in the United States, according to the World Health Organization. In a recent study, researchers found that almost half of college students and their non-college-attending peers had met the criteria for a diagnosable mental illness in the previous year, with alcohol use disorders the most common. And yet much less is known about the potential risk factors of mental health problems for young adults than almost any other age group.

Abnormalities in white matter tracts involved in attention may also explain the frequency of adult ADHD, which usually first appears in childhood or adolescence but may have gone undiagnosed. From a quality-of-life standpoint, young adults with ADHD suffer considerable disruption of their daily lives. Studies have shown they achieve fewer educational goals beyond high school than young adults without ADHD symptoms, are less likely to be employed full-time, and have lower average household incomes. Other studies have found they are also twice as likely to get arrested or divorced, 78 percent more likely to be addicted to cigarettes or cigars, three times more likely to be out of work, and four times more likely to contract a sexually transmitted disease.

How vulnerable are emerging adults? While juvenile crime has actually decreased, the incarceration rates for young adults, ages twenty to twenty-four, nearly doubled over the last decade. Young adults with mental health issues are at higher risk of dropping out of college, having unplanned pregnancies, being unemployed, and suffering from drug and alcohol abuse. And yet the drop-off in mental health services from teenage years to young adulthood is drastic. More than twice as many teens as emerging adults receive inpatient or residential care. And this doesn't include the fact that today we have a whole generation of war veterans in their twenties and thirties with special needs and problems, both physical *and* mental.

Young adults have not been forsaken, however. There are residential treatment centers for emerging adults like New Lifestyles in Winchester, Virginia; websites, like the Network on Transitions to Adulthood (transitions2adulthood.com), that address the specific concerns of postadolescent young adults; and books like *Coming of Age in America: The Transition to Adulthood in the Twenty-First Century*. There is even a Society for the Study of Emerging Adulthood and a scholarly journal dedicated to this discrete age group.

With the neuroscience of young adulthood still in its infancy, we don't yet know if the twenties represent a last chance to capitalize on the final stage of brain development; whether, as parents or scientists or educators, we should be urging twentysomethings to find a skilled job—any job—before their learning curves begin their slow dive with aging or whether we should urge the opposite, to take a final fling at exploring possibilities while creative thinking is still at its neurobiological height. Jay Giedd, the NIMH scientist who has been studying the teenage brain longer than almost anyone else, told a reporter, "It's too soon to tell." But if we've learned anything about neuroscience in the past decade, then maybe the answer should be "Stay tuned."

Postscript
Final Thoughts

At the end of the day—and I know the days can be very, very long when you have a teenager—your kids still have to grow up and develop and learn and mature on their own. You can't do it for them. Hopefully this book has helped you understand that your teenagers are not aliens, they are not another species from another planet, but they are at a critical stage in their development where everything is not yet completely in sync, and the more you understand this—and can explain it to them—the smoother these years will be. They will never be completely smooth, of course, and there will always be times when you want to pull your hair out. So here are a few important reminders:

- Be tolerant of your teens' misadventures, but make sure you talk to them calmly about their mistakes.
- Don't be shocked when your teens do something stupid and then say they don't know why. *You* now know why, but explain that to them—how their prefrontal lobes haven't quite come

online yet. And remember, even the smartest, most obedient, meekest kids will do something stupid before "graduating" from adolescence.

- Communicate and relate: Emphasize the positive things in your teens' lives and encourage them to try different activities and new ways of thinking about things. Reinforce that you are there for them when they need advice.
- Social networking tools and websites are an important avenue of communication with your teens. Some parents report that their most successful and meaningful "conversations" with their teens occurred while texting back and forth with them. And if you don't know how to text yet, ask your teenager.

I hope that this book will give you many pieces of information to consider and, when appropriate, to use to start conversations with your teen. As I said in the beginning, teenagers respect information and are also naturally curious as to who they are. You can also use the illustrations in this book to help them understand what a special place they are in.

Ideally, conversations about facts—such as those contained in this book—will help you to avoid confrontations with your teens and prevent getting into oppositional patterns with them. This is a time of learning, so appeal to that whenever possible. Judgments handed out in anger or without an explanation only increase the alienation of a teen. Criticisms of behavior are best when they are followed by a "because." The behavior has to be taken in context, too. A kid who is a straight-A student who does homework while standing on his head is not a problem, even if the parent feels this is an unconventional approach! However, if your teenager is struggling with low grades and organizational issues and is progressively falling behind, it is your job to step in to offer help and brainstorm why these things are happening. Although you may think you have a quasi-independent young person in your household, especially compared with the child he or she once was, you will have to invest more time and effort in your teen than you probably ever anticipated. You are the parent. You need to try to elicit from your teens whether they even care that they're getting bad grades. If they don't care, then that's a much deeper problem than how they study. At that point you need to stop and try to determine if this is just an assertion of independence and control. If so, then why does the child feel he or she needs to demonstrate control? Are there crises in the household or the teens' lives that are troubling them? Are they being peer-pressured to be in a group that prides itself on academic indifference? Are they abusing drugs? Or do they have self-esteem issues that might be an early sign of depression or other mental illness? All these entail different action plans.

On the other hand, if they themselves are frustrated, they are likely to be at least somewhat receptive to some gentle problem-solving advice from you, the parent. Start by asking them if they think their haphazard study situation is effective. Coax an answer out of them if at all possible—this will help them exercise their own ability to solve problems independently. As a last resort, suggest "directions" as to what might work, and if your kid at least tries to follow through, then you've at least set up a situation of positive reinforcement for future use. Offering some sort of minimal reward will help, too, as it is unlikely grades will turn around overnight and be their own reward.

You want to always remain as positive as you can because you want to empower your teenagers and help them understand what an amazing time of their lives this is, a time of opportunity. Your job isn't to stifle them but rather to help them channel their energies in positive directions. One major way you can facilitate this is by providing a calm and organized environment in which they can grow. The less stressful and chaotic *your* life is, the less stressful and chaotic theirs will be.

You also really want to be aware of the dangers out there and what your teenagers are being exposed to, so you need to immerse yourself in their world. Learn what music they listen to, what TV shows and movies they watch, and what books they are reading. You don't need to be your teen's BFF (best friend forever); you just need to know what's going on in his or her life so you can better understand, advise, and set limits.

Ultimately, you are your child's first and most important role model. Your children are watching you, even though they may not even be conscious of it. How you approach your own life, how you confront your own challenges, provides learning experiences for them, so share it with them without overwhelming them. You are a team, after all. That really hit home for me in 2000 when my two sons and I went through the trauma of a house fire. Both boys were home when I arrived from work at around six o'clock. Andrew had a wrestling match the next day and had lost his shoes, so we got back in the car—all three of us—to buy another pair. We were gone only about thirty minutes when my beeper went off. It was the alarm system from the house. So I turned the Chrysler minivan around, and as we got closer to the house, I could see the fire trucks. It was shocking at first. Then I remember thinking, *It's just "stuff," whatever we've lost, because I have everything I need right here: my two boys are safe in the back of the car.* That's all that mattered.

I remember that as we walked through the crowd that had gathered around our burning house, I watched my kids and saw how my neighbors just surrounded them, comforting them. The house looked okay from the outside, but inside it was gutted. Eventually we were told an electric can opener had tipped over and somehow ignited some Tupperware containers. I never let my sons back into the house until it was nearly restored because I didn't want them to have that image of their burned bedrooms. That's the kind of stress children and teenagers don't need, and at least I could protect them from that. My neighbors knew; that's why they surrounded the boys and whisked them away from the scene.

In the months that followed I realized that as traumatic as the fire was, it was a valuable bonding experience for us, too. My sons learned that material things don't matter. I learned that it could have been so much worse. Together we were a team, and I kept telling them that: that it was okay, that we were here, alive, and we would make a fresh start. That's all that mattered. At that time Andrew was thirteen and Will was eleven. Teen times were just starting. This was only the first of the challenges we would face, but we would face them together.

Acknowledgments

I would like to thank my coauthor, Amy Ellis Nutt, for her patience and support throughout the process of writing this book. Amy's zeal for knowledge and investigative know-how helped glue together the many seemingly fragmented pieces of information we present here. The process of writing this book together produced another result—the discovery of a new friend and a friendship that will outlast this book and its future editions!

Words cannot describe my gratitude to my close friends and family, especially my sons, Andrew and Will; my parents; and of course Jeff for all those times they put up with postponed weekend events and other activities because I was "doing my book." I have decided Jeff is the most patient person on the planet and deserves special thanks for all his support.

Much appreciation to Marcus Handy, a research specialist, and Mary Leonard, a biomedical artist, both at the University of Pennsylvania, for researching and producing, respectively, the final set of illustrations used in this book.

I would like to acknowledge many colleagues both at Harvard and at the University of Pennsylvania who have encouraged me to keep at the writing, and who have shared their research with me as I wrote this book. For the earliest encouragement, I would like to thank Patty Hager, former dean of students at Concord Academy, Massachusetts, whose many questions prompted me to respond by creating a program for the Elizabeth Hall symposium. I would like to thank my colleagues Dr. David Urion, associate professor of neurology and associate attending at Boston Children's Hospital, and Maryanne Wolf, PhD, professor of neuropsychology at Tufts University in Massachusetts—both David and Maryanne shared the stage with me on several occasions as we gave early versions of our "Teen Brain 101" talks.

And last but definitely not least, thanks also go to Wendy Strothman for her advocacy of the need for this book, and to our editor, Claire Wachtel, for her wise counsel and clear thinking on and off its pages.

Glossary

action potential: pulse of electrical activity that occurs when a neuron sends information down its axon.

amygdala: almond-shaped structure deep within the brain, at the front of the temporal lobe; the amygdala is where emotions are processed.

arborization: growth and branching out of dendrites

astrocytes: star-shaped cells in the brain and spinal cord. They are the most abundant cells in the brain and are responsible for a wide variety of functions, such as providing nutrients to neural tissue and helping repair cells in the brain and spinal cord damaged by traumatic injury.

axon: part of the neuron that carries an impulse toward the synapse, which then transmits a chemical message to other neurons.

brain plasticity: the capacity of the brain to change structure and function, especially in response to the stimulation of repeated experiences. Plasticity is at its highest during childhood, when the brain is developing, but it is also present during adulthood, and it is the mechanism for learning and memory.

cannabinoids: both a class of chemical compounds that are found within the cannabis plant (marijuana) and also forms that occur naturally in the brain (called endocannabinoids). These substances interact with specific endogenous receptors in the brain that help reduce pain and anxiety.

cortisol: a hormone produced in the human body by the adrenal gland and released in response to stress, diverting energy toward any immediate threat, small or large. Many things in the environment can affect cortisol levels, from caffeine to sleep deprivation. Too much cortisol can suppress the immune system, raise blood pressure, and damage neurons, especially in the hippocampus.

dendrite: the branch-like part of a neuron that receives information from other neurons. Dendrites make contacts with the axons of nearby neurons, and usually the synapses are found on dendritic spines.

dopamine: a neurotransmitter that helps control the brain's reward and pleasure centers.

endocannabinoids: the brain's own naturally occurring cannabinoids, composed of molecules known as lipids.

endorphins: produced by the body during exercise, they interact with opiate receptors in the brain to reduce our perception of pain, much like morphine.

epinephrine: also known as adrenaline, a hormone that is released in response to strong emotions, such as fear and anger, and that, by sending energy to the muscles and increasing the heart rate, prepares the body for "fight or flight."

frontal lobe: one of the four main regions of the brain, positioned at the frontmost portion of the brain and responsible for many executive functions, such as reasoning, planning, and other complex cognitive processes.

GABA: acronym for gamma-aminobutyric acid, a neurotransmitter that inhibits activity at the synapse. It's important if your brain is overreacting to a stimulus.

glia: helper cells that aid myelin production, for instance, or that, in the case of astrocytes, a type of glial cell, digest parts of dead neurons. Unlike neurons, glial cells do not carry nerve impulses.

glutamate: synaptic receptors that are the brain's main excitatory neurotransmitters. They are particularly important in memory formation and learning.

gray matter: the cortex, or part of the brain that contains neurons, as opposed to the white matter, which underlies the cortical gray matter.

gyri: the ridges in the brain's wrinkled gray matter.

hippocampus: the part of the brain, located in the temporal lobes, involved chiefly in the processing of memories.

hypothalamus: the part of the brain, located just above the brainstem, that is responsible for major metabolic processes such as body temperature, hunger, thirst, and sleep.

limbic system: the part of the brain, located on both sides of the thalamus and just under the cerebrum, that includes structures involved in emotion, motivation, and memory.

long-term potentiation: abbreviated LTP, the long-lasting form of plasticity, or enhancement of synaptic transmissions between neurons, that is the main mechanism for learning and memory.

melatonin: hormone produced by the brain's pineal gland that helps regulate circadian sleep rhythms.

myelin: the white matter that lines the axons, helping transmit the brain's signals faster and more efficiently. Myelin covers many kinds of axons in the brain, spinal cord, and peripheral nerves.

neuron: the unique cells found in the nervous system that signal one another using neurotransmitters.

neurotransmitters: the chemical messengers released from one neuron, across the synaptic cleft, to a receptor site on the adjacent neuron.

occipital lobes: the part of the brain, located in the back, that houses the visual cortex.

parietal lobes: the brain region, located behind the frontal lobes and in front of the occipital, that is responsible for various senses, such as touch, as well as for visuospatial processing.

prefrontal lobes: the most anterior part of the frontal lobes, located behind the forehead, that is responsible for, among many other tasks, modulating social behavior, decision-making, and personality expression.

sulci: the grooves on the surface of the cerebral cortex that make up part of the "wrinkles" of the brain's gray matter.

synapse: structure at the end of a neuron through which an electrical or chemical message is passed to another neuron. All thinking, feeling, movement, etc., depends on the transmission of messages through synapses.

synaptogenesis: the process of forming new synapses, occurring throughout a person's life span but mostly during critical periods such as infancy, childhood, and adolescence.

temporal lobes: the areas, located at the sides of the brain, that process smell and sound and complex visual-recognition processes.

THC: acronym for tetrahydrocannabinol, the main active, mind-altering ingredient in cannabis.

ventral tegmental area (VTA): a group of neurons that is close to the midline on the floor of the midbrain, whose neurons connect to many areas of the brain, and that is the origin of the brain's dopamine system. For that reason it plays a large role in the brain's natural reward circuitry and is implicated in drug addiction.

white matter: the part of the brain that contains myelinated axon tracts responsible for transmitting electrical and chemical messages between neurons.

Notes

INTRODUCTION: BEING TEEN

6 *"These years are the best decade of life"*: Granville Stanley Hall, *Adolescence: Its Psychology and Its Relations to Physiology, Anthropology, Sociology, Sex, Crime, Religion and Education* (New York: D. Appleton, 1904).

6 *"the birthday of the imagination"*: Ibid.

CHAPTER 1. ENTERING THE TEEN YEARS

16 *In fact the word "teenager"*: *Popular Science*, Apr. 1941.

16 *With the onset of the Depression*: Sharron Solomon-McCarthy, "The History of Child Labor in the United States: *Hammer v. Dagenhart*," in *The Supreme Court in American Political History* (New Haven: Yale–New Haven Teachers Institute, 2004).

17 *the number one hallmark of adulthood*: Tom W. Smith, "Coming of Age in 21st Century America: Public Attitudes Towards the Importance and Timing of Transitions to Adulthood," National Opinion Research Center, University of Chicago, GSS Topical Report No. 35, Mar. 2003.

17 *"Young people became teenagers"*: Thomas Hine, *The Rise and Fall of the American Teenager* (New York: William Morrow, 1999).

18 *"Character and personality are taking form"*: Granville Stanley Hall, *Adolescence*.

21 *researchers discovered that puberty*: B. B. Van Bockstaele, "Genes Have Been Discovered for the Brain Pathway That Triggers Puberty," *Digital Journal*, Dec. 12, 2008.

22 *adolescence is a time of increased response*: Hui Shen, Qi Hva Gong, et al., "Reversal of Neurosteroid Effects at α4β2δ $GABA_A$ Receptors Triggers Anxiety at Puberty," *Nature Neuroscience* 10, no. 4 (Apr. 2007).

CHAPTER 2. BUILDING A BRAIN

29 *the brain of Albert Einstein*: Sandra F. Witelson, Debra L. Kigar, and Thomas Harvey, "The Exceptional Brain of Albert Einstein," *Lancet* 353, no. 9170 (June 19, 1999).

31 *early-twentieth-century Canadian neuroscientist*: Wilder Penfield and Edwin Boldrey, "Somatic Motor and Sensory Representation in the Cerebral Cortex of Man as Studied by Electrical Stimulation," *Brain* 60, no. 4, (Dec. 1937).

33 *Some of the most famous work*: David H. Hubel and Thorsten N. Wiesel, "The Period of Susceptibility to the Physiological Effects of Unilateral Eye Closure in Kittens," *Journal of Physiology* 206, no. 2 (Feb. 1970).

37 *major study to examine how brain regions*: National Institute of Mental Health, "Teenage Brain: A Work in Progress," NIMH Fact Sheet, 2001. Also, R. K. Lenroot and J. N. Giedd, "Brain Development in Children and Adolescents: Insights from Anatomical Magnetic Resonance Imaging," *Neuroscience and Biobehavioral Reviews* 30, no. 6 (2006).

40 *The parietal lobes help the frontal lobes to focus*: Frederik Edin, Torkel Klingberg, et al., "Mechanism for Top-Down Control of Working Memory," *Proceedings of the National Academy of Sciences* 106, no. 16 (Apr. 3, 2009).

42 *of the nearly six thousand adolescents who die*: Allstate/Sperling's Best Places, "America's Teen Driving Hotspots Study" (May 2008).

42 *Missouri scientists discovered that simultaneous tasks*: Moshe Naveh-Benjamin, Angela Kilb, and Tyler Fisher, "Concurrent Task Effects on Memory Encoding and Retrieval: Further Support for Asymmetry," *Memory and Cognition* 34, no. 1 (Jan. 2006).

45 *connection of the hippocampus to memory*: Luke Dittrich, "The Brain That Changed Everything," *Esquire*, Oct. 25, 2010.

CHAPTER 3. UNDER THE MICROSCOPE

49 *"All infants are born in a state of psychedelic splendor"*: Kathleen McAuliffe, "Life of Brain," *Discover*, June 2007.

57 *Thus, cells that "fire" together*: Carla Shatz et al., "Dendritic Growth and Remodeling of Cat Retinal Ganglion Cells During Fetal and Postnatal Development," *Journal of Neuroscience* 8, no. 11 (Nov. 1988).

58 *Jay Giedd and colleagues*: R. K. Lenroot and J. N. Giedd, "Brain Development in Children and Adolescents: Insights from Anatomical Magnetic Resonance Imaging," *Neuroscience and Biobehavioral Reviews* 30, no. 6 (2006).

60 *Dan Gordon, a fifteen-year-old boy*: Amelia Hill, "Red Cross Study Reveals Problems with Teenagers and Drink," *Guardian*, Sept. 12, 2010.

63 *Bennett Barber was sixteen years old*: Alan Burke, "Cops: Freezing Teen Hit; Friends Lied," *Salem News*, Jan. 23, 2009.

CHAPTER 4. LEARNING: A JOB FOR THE TEEN BRAIN

67 *Before it closed in 1992*: "Massachusetts Gaining in Its Care for the Retarded," *New York Times*, Jan. 4, 1987.

69 *British physiologist and Nobel Prize winner Charles Sherrington*: Charles Sherrington, *Man on His Nature*, reissue edition (Cambridge: Cambridge University Press, 2009).

69 *Donald Hebb, an American neuropsychologist*: Mark A. Gluck et al., *Learning and Memory: From Brain to Behavior* (New York: Worth Publishers, 2007).

70 *young brains are shaped by experience*: Nico Spinelli and Frances Jensen, "Plasticity: the Mirror of Experience," *Science* 203, no. 4375 (Jan. 1979).

70 *cab drivers in London*: Eleanor Maguire et al., "London Taxi Drivers and Bus Drivers: A Structural MRI and Neuropsychological Analysis," *Hippocampus* 16, no. 12 (2006).

71 *learning the tango*: Patricia McKinley et al., "Effect of a Community-Based Argentine Tango Dance Program on Functional Balance and Confidence in Older Adults," *Journal of Aging and Physical Activity* 16, no. 4 (Oct. 2008).

76 *"long periods of excess use or disuse"*: Tim Bliss et al., *Long-Term Potentiation: Enhancing Neuroscience for 30 Years* (Oxford: Oxford University Press, 2004).

77 *effective pruning increased brain efficiency*: Emily Kilroy et al., "Relationships Between Cerebral Blood Flow and IQ in Typically Developing Children and Adolescents," *Journal of Cognitive Science* 12, no. 2 (2011).

79 *your IQ can change during your teen years*: Carol K. Seligman and Elizabeth A. Rider, *Life Span Human Development*, 7th ed. (Belmont, CA: Wadsworth Publishing, 2012). Also, Sue Ramsden et al., "Verbal and Non-Verbal Intelligence Changes in the Teenage Brain," *Nature* 479, no. 7371 (Oct. 19, 2011).

80 *children with high IQs may have an extended learning period*: Angela M. Brant, John K. Hewitt, et al., "The Nature and Nurture of High IQ: An Extended Sensitive Period for Intellectual Development," *Psychological Science* 24, no. 8 (Aug. 2013).

84 *the odds that certain bad things*: Christina Moutsiana, Tali Sharot, et al., "Human Development of the Ability to Learn from Bad News," *Proceedings of the National Academy of Sciences* 110, no. 41 (Oct. 8, 2013).

CHAPTER 5. SLEEP

87 *Infants and children are "larks"*: Jim Horne, *Sleepfaring: A Journey Through the Science of Sleep* (Oxford: Oxford University Press, 2007).

87 *Teenagers can be, and are, forced to abide*: M. H. Hagenauer et al., "The Neuroendocrine Control of the Circadian System: Asociescent Chronotype," *Frontiers in Neuroendocrinology* 33 (2012).

91 *basic cognitive tests in order to fatigue their brains*: Marc G. Berman et al., "The Cognitive Benefits of Interacting with Nature," *Psychological Science* 19, no. 12 (2008).

92 *consolidation of memories happens in two stages*: Robert Stickgold, "Sleep-Dependent Memory Consolidation," *Nature* 437, no. 7063 (Oct. 27, 2005).

94 *relationship between sleep and learning in adolescents*: Edward B. O'Malley and Mary B. O'Malley, "School Start Time and Its Impact on Learning and Behavior," in *Sleep and Psychiatric Disorders in Children and Adolescents*, ed. A. Ivanenko (New York: Informa Healthcare, 2008).

95 *the sleep-learning connection from the opposite direction*: Jeffrey M. Donlea et al., "Use-Dependent Plasticity in Clock Neurons Regulates Sleep Need in *Drosophila*," *Science* 324, no. 5923 (Apr. 3, 2009).

96 *two and a half times more likely to report suicidal thoughts*: Norihito Oshima et al., "The Suicidal Feelings, Self-Injury, and Mobile Phone Use After Lights Out in Adolescents," *Journal of Pediatric Psychiatry* 37, no. 9 (Oct. 2012).

97 *Poor sleep habits may even have a role in juvenile delinquency*: Samantha S. Clinkinbeard et al., "Sleep and Delinquency: Does the Amount of Sleep Matter?" *Journal of Youth and Adolescence* 40, no. 7 (July 2011).

99 *energy-drink-related ER visits increased tenfold*: "Update on Emergency Department Visits Involving Energy Drinks: A Continuing Public Health Concern," SAMHSA *Dawn Report*, Drug Abuse Warning Network, Jan. 10, 2013.

100 *the bright LED light of a computer screen*: Mariana Figueiro et al., "Light Level and Duration of Exposure Determine the Impact of Self-Luminous Tablets on Melatonin Suppression," *Applied Ergonomics* 44, no. 2 (Mar. 2013).

101 *A blue light in LEDs*: Katie Worth, "Casting Light on Astronaut Insomnia: ISS to Get Sleep-Promoting Lightbulbs," *Scientific American*, Dec. 4, 2012.

CHAPTER 6. TAKING RISKS

103 *National Public Radio in March 2010*: Richard Knox, "The Teen Brain: It's Just Not Grown Up Yet," National Public Radio, Mar. 1, 2010, http://www.npr.org/templates/story/story.php?stryId=124119468.

105 *Even Aristotle weighed in*: Aristotle, *The Rhetoric of Aristotle* (London and New York: Macmillan, 1886).

105 *young people were slaves to their passions*: Ibid.

106 *ability of subjects to inhibit their eye movements*: Beatriz Luna et al., "What Has fMRI Told Us About the Development of Cognitive Control Through Adolescence?" *Brain and Cognition* 72, no. 1 (Feb. 2010).

106 *adolescents use a more limited brain region*: Valerie Reyna and Frank Farley, "Risk and Rationality in Adolescent Decision Making: Implications for Theory, Practice and Public Policy," *Psychological Science in the Public Interest* 7, no. 1 (Sept. 2006).

107 *cingulate cortex lights up*: Laurence Steinberg, "A Social Neuroscience Perspective on Adolescent Risk-Taking," *Developmental Review* 28, no. 1 (Mar. 2008).

110 *brain activity of subjects who were asked to make financial decisions*: Brian Knutson et al., "The Neural Basis of Financial Risk-Taking," *Neuron* 47, no. 5 (Sept. 1, 2005).

110 *cards depicting either happy faces or calm faces*: Leah H. Somerville et al., "Frontostriatal Maturation Predicts Cognitive Control Failure to Appetitive Cues in Adolescents," *Journal of Cognitive Neuroscience* 23, no. 9 (Sept. 2011).

111 *"Milton Academy Rocked by Expulsions"*: Michael Levenson and Jenna Russell, "Milton Academy Rocked by Expulsions," *Boston Globe*, Feb. 20, 2005.

112 *teenagers today no longer regard oral sex*: Abigail Jones and Marissa Miley, *Restless Virgins: Love, Sex, and Survival at a New England Prep School* (New York: William Morrow, 2008).

113 *"didn't get HPV, herpes, chlamydia, or HIV"*: Ibid.

113 *youth decision-making questionnaire*: Margo Gardner and Laurence Steinberg, "Peer Influence on Risk-Taking, Risk Preference, and Risky Decision Making in Adolescence and Adulthood: An Experimental Study," *Developmental Psychology* 41, no. 4 (July 2005).

CHAPTER 7. TOBACCO

118 *90 percent of new smokers*: Regina Benjamin, "Preventing Tobacco Use Among Youth and Young Adults: A Report of the Surgeon General, 2012," http://www.surgeongeneral.gov/library/reports/preventing-youth-tobacco-use/index.html.

118 *connection between smoking and lower IQ*: Mark Weiser et al., "Cognitive Test Scores in Male Adolescent Cigarette Smokers Compared to Non-Smokers: A Population-Based Study," *Addiction* 105, no. 2 (Feb. 2010).

118 *children who are routinely exposed to secondhand smoke*: Kimberly Yolton, Richard Hornung, et al., "Exposure to Environmental Tobacco Smoke and Cognitive Abilities Among US Children and Adolescents," *Environmental Health Perspectives* 113, no. 1 (Jan. 2005).

120 *a cigarette a month for an adolescent*: Joseph DiFranza et al., "Symptoms of Tobacco Dependence After Brief, Intermittent Use: The Development and Assessment of Nicotine Dependence in Youth-2 Study," *Archives of Pediatric and Adolescent Medicine* 161, no. 7 (July 2007).

120 *"when you first get addicted"*: Brenda Wilson, "Study: A Cigarette a Month Can Get a Kid Hooked," National Public Radio, May 31, 2010, http://www.npr.org/templates/story/story.php?storyId=127241145.

122 *single day of smoking cigarettes in adolescence*: Sergio D. Iniguez et al., "Nicotine Exposure During Adolescence Induces a Depression-Like State in Adulthood," *Neuropsychopharmacology* 34, no. 6 (May 2009).

CHAPTER 8. ALCOHOL

126 *inside the* Milford Daily News: Heather McCarron, "Taylor Meyer Laid to Rest," *Milford Daily News*, Oct. 29, 2008.

127 *"If recreational drugs were tools"*: Aaron White, "What Happened? Alcohol, Memory Blackouts, and the Brain," National Institute on Alcohol Abuse and Alcoholism, 2004.

127 *teenagers who watch PG-rated movies*: Susanne E. Tanski, James D. Sargent, et al., "Parental R-Rated Movie Restriction and Early-Onset Alcohol Use," *Journal of Studies on Alcohol and Drugs* 71, no. 3 (May 2010).

127 *In France, where the minimum drinking age*: Kim Willsher, "Lyon Aims to Reduce Le Binge Drinking," *Guardian*, July 17, 2011.

128 *college students tend to pattern their drinking*: H. Wesley Perkins et al., "Misperceptions of the Norms for the Frequency of Alcohol and Other Drug Use on College Campuses," *Journal of American College Health* 47, no. 6 (May 1999).

132 *When rapid or binge drinking results in a blackout*: Michael A. Taffe et al., "Long-Lasting Reduction in Hippocampal Neurogenesis by Alcohol Consumption in Adolescent Nonhuman Primates," *PNAS* 107, no. 24 (June 1, 2010).

134 *An alarming number of college students*: Patrick M. O'Malley et al., "Epidemiology of Alcohol and Other Drug Use Among American College Students," *Journal of Studies on Alcohol* 14, supplement (Mar. 2002).

135 *Human studies of binge-drinking adolescents*: Susan F. Tapert and Sunita Bava, "Adolescent Brain Development and the Risk for Alcohol and Other Drug Problems," *Neuropsychology Review* 20, no. 4 (Dec. 2010).

137 *American Academy of Pediatrics finally published a policy statement*: American Academy of Pediatrics Committee on Substance Abuse, "Alcohol Use by Youth and Adolescents: A Pediatric Concern," *Pediatrics* 125, no. 5 (May 1, 2010).

138 *Those who are monitored closely by their parents*: Caitlin Abar and Robert Turrisi, "How Important Are Parents During the College Years? A

Longitudinal Perspective of *Indirect* Influence Parents Yield on Their College Teens' Alcohol Use," *Addiction Behavior* 33, no. 10 (Oct. 2008).

138 *"the more teenagers drink at home"*: Haske van der Vorst et al., "Do Parents and Best Friends Influence the Normative Increase in Adolescents' Alcohol Use at Home and Outside the Home?" *Journal of Studies on Alcohol and Drugs* 71, no. 1 (Jan. 2010).

140 *A month after the Massachusetts teen's death*: Heather McCarron, "Arrested Teens Accused of 'Hypocrisy,'" *Milford Daily News*, Nov. 25, 2008.

CHAPTER 9. POT

142 *Pot is now regarded by many experts*: Lynn Fiellin et al., "Previous Use of Alcohol, Cigarettes, and Marijuana and Subsequent Use of Prescription Opioids in Young Adults," *Journal of Adolescent Health* 52, no. 2 (Feb. 2013).

142 *marijuana remains the most popular*: United Nations Office on Drugs and Crime, *World Drug Report, 2013* (New York: United Nations, 2013).

144 *Anslinger testified before Congress*: Martin Booth, *Cannabis: A History* (Great Britain: Doubleday, 2003).

144 *no association between pot smoking*: Rufus King and James T. McDonough Jr., "Anslinger, Harry Jacob, and U.S. Drug Policy," in *Encyclopedia of Drugs, Alcohol, and Addictive Behavior*, ed. Rosalyn Carson-Dewitt (New York: Macmillan, 2001).

146 *The first major research breakthrough*: William A. Devane, Allyn C. Howlett, et al., "Determination and Characterization of a Cannabinoid Receptor in Rat Brain," *Molecular Pharmacology* 34, no. 5 (Nov. 1988).

147 *THC affects the suppression of pain*: David Finn et al., "A Role for the Ventral Hippocampal Endocannabinoid System in Fear-Conditioned Analgesia and Fear Responding in the Presence of Nociceptive Tone in Rats," *Pain* 152, no. 11 (Nov. 2011).

149 *In a blog run by the* 420 Times: "Ask an Old Hippie: Help! My Teenage Daughter Is Smoking Marijuana!" *420 Times*, Aug. 19, 2011, http://the420times.com/2011/08/ask-an-old-hippie-help-my-teenage-daughter-is-smoking-marijuana.

152 *the link between chronic pot use*: M. H. Meier et al., "Persistent Cannabis Users Show Neuropsychological Decline from Childhood to Midlife," *Proceedings of the National Academy of Sciences* 109, no. 40 (2012).

152 *schizophrenics have less white matter*: Matthijs G. Bossong et al., "Adolescent Brain Maturation, the Endogenous Cannabinoid System and the Neurobiology of Cannabis-Induced Schizophrenia," *Progress in Neurobiology* 92, no. 3 (Nov. 2010).

152 *In a 2010 article in the* Toronto Star: Nancy J. White, "Marijuana Can Send a Brain to Pot," *Toronto Star*, July 9, 2010.

154 *cannabis use was an important causal factor*: Rebecca Kuepper et al., "Continued Cannabis Use and Risk of Incidence and Persistence of Psychotic Symptoms: 10 Year Follow-up Cohort Study," *British Medical Journal* 342 (Mar. 1, 2011).

155 *early marijuana use was linked to a 50 percent increase*: Ron de Graaf, James C. Anthony, et al., "Early Cannabis Use and Estimated Risk of Later Onset of Depression Spells: Epidemiologic Evidence from the Population-Based World Health Organization World Mental Health Survey Initiative," *American Journal of Epidemiology* 172, no. 2 (July 15, 2010).

157 *losing their parents' trust and respect*: Stephen N. Campbell, "Substance Abuse in Children and Adolescents: Information for Parents and Educators," National Association of School Psychologists, 2004.

CHAPTER 10. HARD-CORE DRUGS

159 *Irma Perez, a fourteen-year-old*: "Irma Perez," A Vigil for Lost Promise, undated, http://www.nationalparentvigil.com/irma.html.

161 *two young people died at New York City's Electric Zoo*: Richard Zitrin et al., "City Cancels Final Day of Electric Zoo Dance Music Festival After Deaths of Two Concertgoers, Possible Sexual Assault," *New York Daily News*, Sept. 1, 2013.

163 *Dopamine concentrations in these two regions*: Heather C. Brenhouse and Susan L. Andersen, "Delayed Extinction and Stronger Reinstatement of Cocaine Conditioned Place Preference in Adolescent Rats, Compared to Adults," *Behavioral Neuroscience* 122, no. 2 (Apr. 2008).

165 *It didn't take long for Ian Eaccarino*: Ginger Katz, "Ian's Story," The Courage to Speak, undated, https://www.couragetospeak.org/AboutUs/CouragetoSpeakStories/IansStory.aspx.

166 *adolescents exposed to OxyContin*: Mary Jeanne Kreek, Yong Zhang, et al., "Behavioral and Neurochemical Changes Induced by Oxycodone Differ Between Adolescent and Adult Mice," *Neuropsychopharmacology* 34, no. 4 (Mar. 2009).

168 *Two eighth-grade girls pleaded guilty*: Michelle Durand, "Teenager in Ecstasy Death Takes Deal," *Daily Journal* (San Mateo, CA), July 8, 2004.

CHAPTER 11. STRESS

173 *Stress in adolescents works differently*: Sheryl S. Smith, "The Influence of Stress at Puberty on Mood and Learning: Role of the α4βδ $GABA_A$ Receptor," *Neuroscience* 249 (Sept. 2012).

175 *when adolescent rats were exposed to social isolation*: Melanie P. Leussis, Susan L. Andersen, et al., "Depressive-Like Behavior in Adolescents After Maternal Separation: Sex Differences, Controllability and GABA," *Developmental Neuroscience* 34, nos. 2–3 (2012).

175 *a quarter of all adolescents by the age of sixteen*: John Fairbank et al., "Building National Capacity for Child and Family Disaster Mental Health Research," *Professional Psychology, Research and Practice* 41, no. 1 (Feb. 1, 2010).

176 *fMRI to show brain activation*: BJ Casey et al., "Biological Substrates of Emotional Reactivity and Regulation in Adolescence

During an Emotional Go-Nogo Task," *Biological Psychiatry* 63, no. 10 (May 15, 2008).

178 *46 percent of those who had been deployed*: Craig Bryan, "Understanding and Preventing Military Suicide," *Archives of Suicide Research* 16, no. 2 (2012).

179 *adolescents had less gray matter in the prefrontal cortex*: Erin Edmiston et al., "Corticostriatal-Limbic Gray Matter Morphology in Adolescents with Self-Reported Exposure to Childhood Maltreatment," *Archives of Pediatric and Adolescent Medicine* 165, no. 12 (Dec. 2011).

180 *number of ways to help adolescents*: American Psychological Association, "Children and Trauma," Presidential Task Force on Posttraumatic Stress Disorder and Trauma in Children and Adolescents, 2008.

CHAPTER 12. MENTAL ILLNESS

186 *Three-quarters of young adults with psychiatric illness*: J. Kim-Cohen, A. Caspi, et al., "Prior Juvenile Diagnoses in Adults with Mental Disorder: Developmental Follow-Back of a Prospective-Longitudinal Cohort," *Archives of General Psychiatry* 60, no. 7 (July 2003).

186 *Also, the more minor problem of adolescent conduct disorder (CD) and oppositional defiant disorder (ODD)*: Kathleen R. Merikangas, Ronald C. Kessler, et al., "The National Comorbidity Survey Adolescent Supplement (NCS-A): I. Background and Measures," *Journal of the American Academy of Child and Adolescent Psychiatry* 48, no. 4 (Apr. 2009).

188 *annual medical costs for a child or teen with a conduct disorder*: Renee Hsia and Myron Belfer, "A Framework for the Economic Analysis of Child and Adolescent Mental Disorders," *International Review of Psychiatry* 20, no. 3 (June 2008).

191 *Adolescent depression is more likely to be chronic*: Centers for Disease Control and Prevention, "Suicide Prevention: Youth Suicide,"

Jan. 2014, http://www.cdc.gov/violenceprevention/pub/youth_suicide.html.

192 *"black-box" label warning*: National Institute of Mental Health, "Anti-Depressant Medications for Children and Adolescents: Information for Parents and Caregivers," undated, http://www.nimh.nih.gov/health/topics/child-and-adolescent-mental-health/antidepressant-medications-for-children-and-adolescents-information-for-parents-and-caregivers.shtml.

194 *the day before she took her own life*: Deborah Sontag, "Who Was Responsible for Elizabeth Shin?" *New York Times*, Apr. 28, 2002.

197 *Calen also spoke at the conference*: Calen Pick, "Bringing Change to Mind on Mental Illness," Society for Neuroscience annual meeting, San Diego, Nov. 15, 2010.

199 *use of cannabis in the early teens can hasten*: Helene Verdoux, "Cannabis Use and Psychosis: A Longitudinal Population-Based Study," *American Journal of Epidemiology* 156, no. 4 (Apr. 17, 2002).

203 *girls show more brain activity*: National Institute of Mental Health, "Brain Emotion Circuit Sparks as Teen Girls Size Up Peers," July 15, 2009, http://www.nimh.nih.gov/news/science-news/2009/brain-emotion-circuit-sparks-as-teen-girls-size-up-peers.shtml.

203 *the Adolescent Mental Health Cohort*: Sari Frojd et al., "Associations of Social Phobia and General Anxiety with Alcohol and Drug Use in a Community Sample of Adolescents," *Alcohol and Alcoholism* 46, no. 2 (Mar.–Apr. 2011).

CHAPTER 13. THE DIGITAL INVASION OF THE TEENAGE BRAIN

207 *go without their digital tools and toys*: Susan Moeller, "24 Hours: Unplugged," International Center for Media and the Public Agenda and the Salzburg Academy on Media & Global Change, 2011.

208 *one thousand students in twelve countries*: Roman Gerodimos, "Going 'Unplugged': Exploring Students' Relationship with the Media and Its Pedagogic Implications," Centre for Excellence in Media Practice, Bournemouth University, Mar. 2011.

210 *95 percent of all young people*: Amanda Lenhart et al., "Social Media and Mobile Internet Use Among Teens and Young Adults," Pew Research Center, Feb. 3, 2010.

211 *the discovery releases a pleasurable rush*: Dave Mosher, "High Wired: Does Addictive Internet Use Restructure the Brain?" *Scientific American*, June 17, 2011.

213 *A study published in China in 2012*: Fuchin Lin, Hao Lei, et al., "Abnormal White Matter Integrity in Adolescents with Internet Addiction Disorder: A Tract-Based Spatial Statistics Study," *PLoS ONE* 7, no. 1 (2012).

213 *Another study, from Korea*: Soon-Beom Hong, Soon-Hyung Yi, et al., "Reduced Orbitofrontal Cortical Thickness in Male Adolescents with Internet Addiction," *Behavioral and Brain Functions* 9, no. 11 (Mar. 2013).

215 *an increase in density in white matter*: Simone Kuhn and Jurgen Gallinat, "Amount of Lifetime Video Gaming Is Positively Associated with Entorhinal, Hippocampal and Occipital Volume," *Molecular Psychiatry* (Aug. 20, 2013).

217 *the link between video games and symptoms of ADHD*: Philip A. Chan and Terry Rabinowitz, "A Cross-Sectional Analysis of Video Games and Attention Deficit Hyperactivity Disorder Symptoms in Adolescents," *Annals of General Psychiatry* 5, no. 16 (2006).

218 *students who watch television while reading*: Zheng Wang and John M. Tchernev, "The 'Myth' of Media Multitasking: Reciprocal Dynamics of Media Multitasking, Personal Needs, and Gratification," *Journal of Communication* 62, no. 3 (June 2012).

218 *Dr. Wang of Ohio State*: Jeff Grabmeir, "Multitasking May Hurt Your Performance but It Makes You Feel Better," Research and Innovation Communications, Ohio State University, Apr. 30, 2012.

219 *eighteen-year-old Oregon man*: Christina Lopez, "Oregon Teen Arrested After Posting 'Drivin Drunk' Facebook Status," ABCNews .go.com, Jan. 4, 2013.

220 *eighteen-year-old Kentucky woman*: Kevin Dolak, "LOL Facebook Post After DUI Accident Lands Woman in Jail," ABCNews.go .com, Sept. 18, 2012.

221 *The girl who had been suspended*: Sulaiman Abdur-Rahman, "4 Adults, 10 Youths Charged in Center City Disturbance," *Philadelphia Inquirer*, Apr. 11, 2013.

221 *The consequences of misuse of digital media*: Amy Ellis Nutt, "Teens Find World of Hurt at Their Fingertips," *Star-Ledger* (Newark, NJ), Sept. 30, 2010.

224 *Silicon Valley executives admitted*: Matt Richtel, "Silicon Valley Says Step Away from the Device," *New York Times*, July 23, 2012.

CHAPTER 14. GENDER MATTERS

229 *connectivity between brain areas in women's versus men's brains*: Racquel E. Gur, Ruben C. Gur, et al., "Sex Differences in the Structural Connectome of the Human Brain," *Proceedings of the National Academy of Sciences* 111, no. 2 (Sept. 2013).

231 *adolescent females have superior language abilities*: James R. Booth et al., "Sex Differences in Neural Processing of Language Among Children," *Neuropsychologia* 46, no. 5 (Mar. 2008).

CHAPTER 15. SPORTS AND CONCUSSIONS

242 *the complex and frightening truth*: Semyon Slobounov et al., "Sports-Related Concussion: Ongoing Debate," *British Journal of Sports Medicine* 48, no. 2 (Jan. 2014).

243 *power generated in athletic collisions*: Steve Broglio, "Biomechanical Properties of Concussions in High School Football," *Medicine and Science in Sports and Exercise* 42, no. 11 (Nov. 2010).

243 *generates just under 3 gs*: Suzanne Slade, *Feel the G's: The Science and Gravity of G-Forces* (Mankato, MN: Compass Point Books, 2009).

244 *brain damage can result even from nonconcussive blows*: Eric Nauman et al., "Functionally-Detected Cognitive Impairment in High

School Football Players Without Clinically-Diagnosed Concussion," *Journal of Neurotrauma* 31, no. 4 (Feb. 15, 2014).

247 *Sarah Ingles, a high school basketball player*: Alan Schwarz, "Girls Are Often Neglected Victims of Concussions," *New York Times*, Oct. 2, 2007.

248 *Second-impact syndrome can also be deadly*: Nadia Kounang, "Brain Bank Examines Athletes' Hard Hits," CNN, Jan. 27, 2012, http://www.cnn.com/2012/01/27/health/big-hits-broken-dreams-brain-bank.

249 *a child's or teenager's brain is not as resilient*: Charles H. Tator, "Sport Concussion Education and Prevention," *Journal of Clinical Sport Psychology* 6, no. 3 (Sept. 2012).

250 *Radiologists at New York University*: "Can Just One Concussion Change the Brain?" National Public Radio, Mar. 15, 2013, http://www.npr.org/2013/03/15/174409382/can-just-one-concussion-change-the-brain.

CHAPTER 16. CRIME AND PUNISHMENT

257 *judge sentenced Terrance Jamar Graham*: Supreme Court of the United States, *Graham v. Florida*, no. 08-7412, argued Nov. 9, 2009, decided May 17, 2010, http://www.supremecourt.gov/opinions/09pdf/08-7412.pdf.

258 *The question the high court was considering*: International Justice Project, "Background—The Constitutionality of the Juvenile Death Penalty," Feb. 12, 2004, http://www.internationaljusticeproject.org/juvConst.cfm.

259 *children have been regarded simply as pint-size adults*: William Blackstone, *Commentaries on the Laws of England*, book IV, ch. 2 (London: Clarendon Press, 1769).

259 *first record of a juvenile being put to death in America*: Eugene Aubrey Stratton, *Plymouth Colony: Its History and People, 1620–1691* (Provo, UT: Ancestry Publishing, 1986).

261 *Hannah Ocuish, a twelve-year-old*: William H. Channing, "A

Sermon, Preached at New-London, December 20th, 1786, Occasioned by the Execution of Hannah Ocuish, a Mulatto Girl, Aged 12 Years and 9 Months, for the Murder of Eunice Bolles, Aged 6 Years and 6 Months" (New London, CT: T. Greene, 1786).

263 *The House of Refuge took charge*: New York City Department of Juvenile Justice, "Juvenile Detention in New York: Then and Now" (a display at John Jay College of Criminal Justice, New York, 1999).

263 *The first true juvenile court*: Julian Mack, "The Juvenile Court," *Harvard Law Review* 23 (1909).

271 *on May 17, 2010, the United States Supreme Court*: Adam Liptak and Ethan Bronner, "Justices Bar Mandatory Life Terms for Juveniles," *New York Times*, June 26, 2012.

276 *"it is virtually impossible to parse the role of the brain"*: Jay N. Giedd, "Adolescent Maturity and the Brain: The Promise and Pitfalls of Neuroscience Research in Adolescent Health Policy," *Journal of Adolescent Health* 45, no. 3 (June 27, 2010).

CHAPTER 17. BEYOND ADOLESCENCE: IT'S NOT OVER YET

280 *continues to be debated by sociologists, psychologists, and scientists*: Kenneth Keniston, "Youth: A 'New' Stage of Life," *American Scholar* 39, no. 4 (Autumn 1970).

282 *Arnett postulated a distinct stage*: Jeffrey Arnett, *Emerging Adulthood: The Winding Road from the Late Teens Through the Twenties* (Oxford: Oxford University Press, 2004).

283 *"Heather is moving to California"*: Lena Dunham, screenwriter, "The Return," *Girls*, HBO, season 1, episode 6, 2012.

284 *the millennial generation's apparent self-centeredness*: Joel Stein, "The New Greatest Generation: Why Millennials Will Save Us All," *Time*, May 20, 2013.

285 *"The developmental agenda for all emerging adults"*: Laura Humphrey, "A Developmental Psycho-Neurobiological Approach to Assessment of Emerging Adults," *Yellowbrick Journal of Emerging Adulthood* 1, no. 1 (2010).

286 *white matter is still being laid down*: Catherine Lebel and Christian Beaulieu, "Longitudinal Development of Human Brain Wiring Continues from Childhood into Adulthood," *Journal of Neuroscience* 31, no. 30 (July 27, 2011).

Selected Bibliography

Acheson, S.; Richardson, R.; and Swartzwelder, H. "Developmental Changes in Seizure Susceptibility During Alcohol Withdrawal." *Alcohol* 18 (1999).

Acheson, S.; Stein, R.; and Swartzwelder, H. "Impairment of Semantic and Figural Memory by Acute Alcohol: Age-Dependent Effects." *Alcoholism: Clinical and Experimental Research* 22 (1998).

Adam, E. "Transactions Among Adolescent Trait and State Emotion and Diurnal and Momentary Cortisol Activity in Naturalistic Settings." *Psychoneuroendocrinology* 31, no. 5 (June 2006).

Adam, E.; Doane, L.; et al. "Prospective Prediction of Major Depressive Disorder from Cortisol Awakening Responses in Adolescence." *Psychoneuroendocrinology* 35, no. 6 (July 2010).

Amnesty International. "Indecent and Internationally Illegal: The Death Penalty Against Child Offenders." AMR 51/143/2002 (2002), http://www.amnesty.org/en/library/asset/AMR51/143/2002/en/060e0781-d7e8-11dd-9df8-936c90684588/amr511432002en.pdf.

Anderson, P.; De Bruijn, A.; Angus, K., et al. "Impact of Alcohol Advertising and Media Exposure on Adolescent Alcohol Use: A Systematic Review of Longitudinal Studies." *Alcohol and Alcoholism* 44, no. 3 (2009).

Andrew, M.; Smith, B.; et al. "The 'Inner Side' of the Transition to Adulthood: How Young Adults See the Process of Becoming an Adult." *Advances in Life Course Research* 11 (Jan. 2006).

Andrews, M. "Why Do We Use Facial Expressions to Convey Emotions?" *Scientific American*, Nov. 8, 2010.

Arnett, J. *Emerging Adulthood: The Winding Road from the Late Teens Through the Twenties*. Oxford: Oxford University Press, 2004.

Arnone, D., et al. "Corpus Callosum Damage in Heavy Marijuana Use: Preliminary Evidence from Diffusion Tensor Tractography and Tract-Based Spatial Statistics." *NeuroImage* 41, no. 3 (July 1, 2008).

Baillargeon, A.; Lassonde, M.; et al. "Neuropsychological and Neurophysiological Assessment of Sport Concussion in Children, Adolescents and Adults." *Brain Injury* 26, no. 3 (2012).

Baird, A., et al. "What Were You Thinking? An fMRI Study of Adolescent Decision-Making." Poster presented at the annual meeting of the Cognitive Neuroscience Society, 2005.

Baird, A., and Fugelsang, J. "The Emergence of Consequential Thought: Evidence from Neuroscience." *Philosophical Transactions of the Royal Society of London, Series B: Biological Sciences* 359, no. 1451 (Nov. 29, 2004).

Baumrind, D., ed. "Why Adolescents Take Chances—and Why They Don't." First commemorative address sponsored by the National Institute of Child Health and Human Development on the occasion of Child Health Day, 1983.

Bawden, D., and Robinson, L. "The Dark Side of Information: Overload, Anxiety and Other Paradoxes and Pathologies." *Journal of Information Science* 35, no. 2 (Apr. 2009).

Bawden, D. and Robinson, L. "A Distant Mirror: The Internet and the Printing Press." *ASLIB Proceedings* 52, no. 2 (2000).

Beckman, M. "Adolescence: Akin to Mental Retardation?" *Science* 305, no. 5684 (July 30, 2004).

Beckman, M. "Crime, Culpability, and the Adolescent Brain." *Science* 305, no. 5684 (July 30, 2004).

Benedict, C. "Mute 19 Years, He Helps Reveal Brain's Mysteries." *New York Times*, July 4, 2006.

Bentley, P. "Is This Proof Smoking Lowers Your IQ? Study Suggests Those on 20 a Day Are Less Intelligent." *Daily Mail Online*, Mar. 30, 2010.

Bjork, J.; Knutson, B.; and Hommer, D. "Incentive-Elicited Brain Activation in Adolescents: Similarities and Differences from Young Adults." *Journal of Neuroscience* 24, no. 8 (Feb. 25, 2004).

Blakemore, S., and Choudhury, S. "Development of the Adolescent Brain: Implications for Executive Function and Social Cognition." *Journal of Child Psychology and Psychiatry* 47, no. 3 (Mar. 2006).

Blum, K. "The Addictive Brain: All Roads Lead to Dopamine." *Collier's* (2012).

Blumberg, H.; Edmiston, E.; et al. "Corticostriatal-Limbic Gray Matter Morphology in Adolescents with Self-Reported Exposure to Childhood Maltreatment." *Archives of Pediatric and Adolescent Medicine* 165, no. 12 (Dec. 2011).

Boden, B.; Mueller, F.; et al. "Catastrophic Head Injuries in High School and College Football Players." *American Journal of Sports Medicine* 35, no. 7 (July 2007).

Boot, W., Gratton, G., et al. "The Effects of Video Game Playing on Attention, Memory, and Executive Control." *Acta Psychologica* 129, no. 3 (Nov. 2008).

Brenhouse, H.; Sonntag, K.; and Andersen, S. "Transient D-1 Dopamine Receptor Expression on Prefrontal Cortex Projection Neurons: Relationship to Enhanced Motivational Salience of Drug Cues in Adolescence." *Journal of Neuroscience* 28, no. 10 (Mar. 5, 2008).

Broglio, S., Zimmerman, J., et al. "Head Impacts During High School Football: A Biomechanical Assessment." *Journal of Athletic Training* 44, no. l4 (July–Aug. 2009).

Bronson, P., and Merryman, A. *NurtureShock: New Thinking About Children*. New York: Twelve, 2009.

Brown, S.; Tapert, S.; Granholm, E.; and Delis, D. "Neurocognitive Functioning of Adolescents: Effects of Protracted Alcohol Use." *Alcoholism: Clinical and Experimental Research* 24, no. 2 (Feb. 2000).

Burman, D.; Bitan, T.; and Booth, J. "Sex Differences in Neural Processing of Language Among Children." *Neuropsychologia* 46, no. 5 (Apr. 2008).

Bushy, D.; Tononi, G.; and Cirelli, C. "Sleep and Synaptic Homeostasis: Structural Evidence in *Drosophila*." *Science* 332, no. 6037 (June 24, 2011).

Byrnes, E.; Johnson, N.; and Carini, L. "Multigenerational Effects of Morphine Exposure on the Mesolimbic Dopamine System." Poster presented at the annual meeting of the Society of Neuroscience, Nov. 14, 2010.

Byrnes, J.; Babb, J.; Scanlan, V.; and Byrnes, E. "Adolescent Opioid Exposure in Female Rats: Transgenerational Effects on Morphine Analgesia and Anxiety-Like Behavior in Adult." *Behavioural Brain Research* 218, no. 1 (Mar. 17, 2011).

Cannon, T.; Heinssen, R.; et al. "Prediction of Psychosis in Youth at High Clinical Risk: A Multisite Longitudinal Study in North America." *Archives of General Psychiatry* 65, no. 1 (Jan. 2008).

Cao, J.; Li, M.; et al. "Gestational Nicotine Exposure Modifies Myelin Gene Expression in the Brains of Adolescent Rats with Sex Differences." *Translational Psychiatry* 3 (Apr. 2013).

Carr, N. *The Shallows: What the Internet Is Doing to Our Brains*. New York: W. W. Norton, 2010.

Carrion, V.; Reiss, A.; et al. "Converging Evidence for Abnormalities of the Prefrontal Cortex and Evaluation of Midsagittal Structures in Pediatric Posttraumatic Stress Disorder Study: An MRI Study." *Psychiatry Research* 172, no. 3 (June 30, 2009).

Carskadon, M. "When Worlds Collide: Adolescent Need for Sleep Versus Societal Demands." In *Adolescent Sleep Needs and School Starting Times*, edited by K. Wahlstrom. Phi Delta Kappa Educational Foundation, 1999.

———. *Adolescent Sleep Patterns: Biological, Social, and Psychological Influences*. Cambridge: Cambridge University Press, 2002.

Casey, B.; Getz, S.; and Galvan, A. "The Adolescent Brain." *Developmental Review* 28 (2008).

Casey, B.; Giedd, J.; and Thomas, K. "Structural and Functional Brain Development and Its Relation to Cognitive Development." *Biological Psychology* 54, nos. 1–3 (Oct. 2000).

Casey, B.; Tottenham, N.; et al. "Transitional and Translational Studies of Risk for Anxiety." *Depression and Anxiety* 28, no. 1 (Jan. 2011).

Caster, J.; Walker, Q.; and Kuhn, C. "Enhanced Behavioral Response to Repeated-Dose Cocaine in Adolescent Rats." *Psychopharmacology* 183, no. 2 (Dec. 2005).

Chambers, R.; Taylor, J.; and Potenza, M. "Developmental Neurocircuitry of Motivation in Adolescence: A Critical Period of Addiction Vulnerability." *American Journal of Psychiatry* 160, no. 6 (June 2003).

Chan, P., and Rabinowitz, T. "A Cross-Sectional Analysis of Video Games and Attention Deficit Hyperactivity Disorder Symptoms in Adolescents." *Annals of General Psychiatry* 5, no. 16 (2006).

Cohen, M.; Tottenham, N.; and Casey, B. "Translational Developmental Studies of Stress on Brain and Behavior: Implications for Adolescent Mental Health and Illness?" *Neuroscience* 26, no. 249 (Sept. 2013).

Collingridge, G.; Isaac, J.; and Wang, Y. "Receptor Trafficking and Synaptic Plasticity." *Nature Reviews Neuroscience* 5, no 12 (Dec. 2004).

Colvin, A.; Mullen, J.; Groh, M.; et al. "The Role of Concussion History and Gender in Recovery from Soccer-Related Concussion." *American Journal of Sports Medicine* 37, no. 9 (Sept. 2009).

Common Sense Media. "Social Media, Social Life: How Teens View Their Digital Lives." A Common Sense Media Research Study, June 26, 2012.

Copeland, W.; Costello, E.; et al. "Posttraumatic Stress Without Trauma in Children." *American Journal of Psychiatry* 167, no. 9 (Sept. 2010).

Covassin, T.; Swanik, C.; and Sachs, M. "Sex Differences and the Incidence of Concussions Among College Athletes." *Journal of Athletic Training* 38, no. 3 (2003).

Craft-Rosenberg, M., and Pehler, S., eds. *Encyclopedia of Family Health*, vol. 2. Thousand Oaks, CA: Sage Publications, 2011.

Cuonotte, D.; Spjiker, S.; et al. "Lasting Synaptic Changes Underlie Attention Deficits Caused by Nicotine Exposure During Adolescence." *Nature Neuroscience* 14, no. 4 (Apr. 2011).

Cyranowski, J.; Frank, E.; Young, E.; and Shear, M. "Adolescent Onset of the Gender Difference in Lifetime Rates of Major Depression: A Theoretical Model." *Archives of General Psychiatry* 57, no. 1 (Jan. 2000).

Dahl, J. "Throwaway Children: Juvenile Justice in Collapse." *Crime Report*, Feb. 9, 2010.

Dahl, R., and Spear, L. P. "Adolescent Brain Development: Vulnerabilities and Opportunities." *Annals of the New York Academy of Sciences* (June 2004).

Dawes, M. A., and Dougherty, D. M. "Adolescent Suicidal Behavior and Substance Abuse: Developmental Mechanisms." *Substance Abuse* 31, no. 2 (Oct. 31, 2008).

Daza-Losada, M.; Rodriguez-Arias, M.; Maldonado, C.; et al. "Behavioural and Neurotoxic Long-Lasting Effects of MDMA Plus Cocaine in Adolescent Mice." *European Journal of Pharmacology* 590, nos. 1–3 (Aug. 20, 2008).

Dean, D., and Webb, C. "Recovering from Information Overload." *McKinsey Quarterly* (Jan. 2011).

De Bellis, M.; Clark, D.; Keshavan, M.; et al. "Hippocampal Volume in Adolescent-Onset Alcohol Use Disorders." *American Journal of Psychiatry* 157, no. 5 (May 2000).

De Bellis, M.; Keshavan, M.; Boring, A.; et al. "Sex Differences in Brain Maturation During Childhood and Adolescence." *Cerebral Cortex* 11, no. 6 (June 2001).

DeGaetano, G. *Parenting Well in the Media Age: Keeping Our Kids Human*. Fawnskin, CA: Personhood Press, 2004.

De Graaf, R., et al. "Early Cannabis Use and Estimated Risk of Later Onset of Depression Spells: Epidemiologic Evidence from the Population-Based World Health Organization World Mental Health Survey Initiative." *American Journal of Epidemiology* 172, no. 2 (July 15, 2010).

De Win, M.; Van den Brink, W.; et al. "Sustained Effects of Ecstasy on the Human Brain: A Prospective Neuroimaging Study in Novel Users." *Brain* 131, no. 11 (Nov. 2008).

Diaz-Arrastia, R.; Agostini, M.; Madden, C.; and Van Ness, P. "Posttraumatic Epilepsy: The Endophenotypes of a Human Model of Epileptogenesis." *Epilepsia* 50, no. 2 (Feb. 2009).

DiFranza, J., et al. "Symptoms of Tobacco Dependence After Brief, Intermittent Use: The Development and Assessment of Nicotine Dependence in Youth-2 Study." *Archives of Pediatric and Adolescent Medicine* 161, no. 7 (July 2007).

Do Couto, B.; Minarro, J.; Aguilar, M.; et al. "Adolescent Preexposure to Ethanol and 3,4-Methylenedioxymethylamphetamine (MDMA) Increases Conditioned Rewarding Effects of MDMA and Drug-Induced Reinstatement." *Addiction Biology* 17, no. 3 (May 2012).

Dokoupil, T. "Is the Web Driving Us Mad?" *Newsweek*, July 9, 2012.

Eagleman, D. "The Brain on Trial." *Atlantic*, July–Aug. 2011.

Estelles, J.; Rodriguez-Arias, M.; Maldonado, C.; et al. "Gestational Exposure to Cocaine Alters Cocaine Reward." *Behavioural Pharmacology* 17, nos. 5–6 (Sept. 2006).

European College of Neuropsychopharmacology. "The Emotional Brain in Youth: Research Suggests How to Diagnose and Treat Mood Disorders in Children and Adolescents." *Science Daily*, Sept. 6, 2011, http://www.sciencedaily.com/releases/2011/09/110904140340.htm.

Evans, B. E.; Greaves-Lord, K.; et al. "The Relation Between Hypothalamic-Pituitary-Adrenal (HPA) Axis Activity and Age of Onset Alcohol Use." *Addiction* 107, no. 2 (Feb. 2012).

Feinstein, S., ed. *Secrets of the Teenage Brain: Research-Based Strategies for Reaching and Teaching Today's Adolescents*. Thousand Oaks, CA: Corwin Press, 2009.

Ferguson, A.; Jimenez, M.; and Jackson, R. "Juvenile False Confessions and Competency to Stand Trial: Implications for Policy Reformation and Research." *New School Psychology Bulletin* 7, no. 1 (2010).

Fisher, P., and Pfeifer, J. "Conceptual and Methodological Issues in Neuroimaging Studies of the Effects of Child Maltreatment." *Archives of Pediatrics and Adolescent Medicine* 165, no. 12 (Dec. 2011).

Foa, E., and Andrews, L. *If Your Adolescent Has an Anxiety Disorder: An Essential Resource for Parents*. Oxford: Oxford University Press, 2006.

Foy, M.; Stanton, M.; Levine, S.; and Thompson, R. "Behavioral Stress Impairs Long-Term Potentiation in Rodent Hippocampus." *Behavioral and Neural Biology* 48, no. 1 (July 1987).

Frantz, K.; O'Dell, L.; and Parsons, L. "Behavioral and Neurochemical Responses to Cocaine in Periadolescent and Adult Rats." *Neuropsychopharmacology* 32, no. 3 (Mar. 2007).

Fried, P.; Watkinson, B.; James, D.; and Gray, R. "Current and Former Marijuana Use: Preliminary Findings of a Longitudinal Study of Effects on IQ in Young Adults." *Canadian Medical Association Journal* 166, no. 7 (Apr. 2, 2002).

Frojd, S.; Ranta, K.; Kaltialo-Heino, R.; and Marttunen, M. "Associations of Social Phobia and General Anxiety with Alcohol and Drug Use in a Community Sample of Adolescents." *Alcohol and Alcoholism* 46, no. 2 (2011).

Frommer, L.; Gurka, K.; Cross, K.; Ingersoll, C.; Comstock, R. D.; and Saliba, S. "Sex Differences in Concussion Symptoms of High School Athletes." *Journal of Athletic Training* 46, no. 1 (Jan.–Feb. 2011).

Furstenberg, F.; Settersten, R.; et al. "Growing Up Is Harder to Do." *Contexts* 3, no. 3 (Aug. 2004).

Fuss, J., and Gass, P. "Endocannabinoids and Voluntary Activity in Mice: Runner's High and Long-Term Consequences in Emotional Behaviors." *Experimental Neurology* 224, no. 1 (July 2010).

Galles, N. "A Primer on Learning: A Brief Introduction from the Neurosciences." *Organisation for Economic Co-Operation and Development* (July 2004).

———. "Adolescent Development of the Reward System." *Frontiers in Human Neuroscience* 4, no. 6 (2010).

Galvan, A. "Neural Plasticity of Development and Learning." *Human Brain Mapping* 31, no. 6 (June 2010).

Galvan, A.; Hare, T.; et al. "Earlier Development of the Accumbens Relative to Orbitofrontal Cortex Might Underlie Risk-Taking Behavior in Adolescents." *Journal of Neuroscience* 26, no. 25 (June 21, 2006).

Gardner, H.; Lawn, N.; Fatovich, D.; and Archer, J. "Acute Hippocampal Sclerosis Following Ecstasy Ingestion." *Neurology* 73, no. 7 (Aug. 18, 2009).

Gardner, M., and Steinberg, L. "Peer Influence on Risk-Taking, Risk Preference, and Risky Decision Making in Adolescence and Adulthood: An Experimental Study." *Developmental Psychology* 41, no. 4 (July 2005).

Garrett, A.; Carrion, V.; Reiss, A.; et al. "fMRI Response to Facial Expression in Adolescent PTSD." Presented at 49th Annual Meeting of the American Academy of Child and Adolescent Psychiatry, San Francisco, CA, October 22–27, 2002.

Gerrard, M.; Gibbons, F.; and Gano, M. "Adolescents' Risk Perceptions and Behavioral Willingness." In *Reducing Adolescent Risk: Toward an Integrated Approach*, edited by D. Romer. Thousand Oaks, CA: Sage Publications, 2005.

Gessel, L.; Fields, S.; Comstock, R.; et al. "Concussions Among United States High School and Collegiate Athletes." *Journal of Athletic Training* 42, no. 4 (Oct.–Dec. 2007).

Giedd, J. "The Teen Brain: Primed to Learn, Primed to Take Risks." *Cerebrum*, Feb. 26, 2009.

Gill, K., and Mizumori, S. "Spatial Learning and the Selectivity of Hippocampal Place Fields: Modulation by Dopamine." In *Hippocampal Place Fields: Relevance to Learning and Memory*, edited by S. Mizumori. Oxford: Oxford University Press, 2008.

Gogtay, N. "Dynamic Mapping of Human Cortical Development During Childhood Through Early Adulthood." *Proceedings of the National Academy of Sciences* 101, no. 21 (May 25, 2004).

Gong, G.; He, Y.; and Evans, A. "Brain Connectivity: Gender Makes a Difference." *Neuroscientist* 17, no. 5 (Oct. 2011).

Gould, T. "Addiction and Cognition." *Addiction Science and Clinical Practice* 5, no. 2 (Dec. 2010).

Grady, M. "Concussion in the Adolescent Athlete." *Current Problems in Pediatric and Adolescent Health Care* 40, no. 7 (Aug. 2010).

Gray, P. "The Dramatic Rise of Anxiety and Depression in Children and Adolescents: Is It Connected to the Decline in Play and Rise in Schooling?" *Psychology Today*, Jan. 26, 2010.

Grier, C.; Terwilliger, R.; Teslovich, T.; Velanova, K.; and Luna, B. "Immaturities in Reward Processing and Its Influence on Inhibitory Control in Adolescence." *Cerebral Cortex* 20, no. 7 (2010).

Guerri, C., and Pascual, M. "Mechanisms Involved in the Neurotoxic, Cognitive, and Neurobehavioral Effects of Alcohol Consumption During Adolescence." *Alcohol* 44, no. 1 (Feb. 2010).

Gulley, J.; Paul, K.; and Cox, C. "Lasting Alterations in Synaptic Transmission and Intrinsic Properties of Rat Prefrontal Cortical Neurons Following Adolescent Exposure to Amphetamines." Poster presented at the Society for Neuroscience, 2010.

Gulley, J., and Stanis, J. "Adaptations in Medial Prefrontal Cortex Function Associated with Amphetamine-Induced Behavioral Sensitization." *Neuroscience* 166, no. 2 (Mar. 17, 2010).

Gurian, M. *Boys and Girls Learn Differently! A Guide for Teachers and Parents*. Hoboken, NJ: Jossey-Bass, 2001.

Hagenauer, M., and Lee, T. "The Neuroendocrine Control of the Circadian System: Adolescent Chronotype." *Frontiers in Neuroendocrinology* 33, no. 3 (Aug. 2012).

Hallowell, E., and Ratey, J. *Delivered from Distraction*. New York: Ballantine Books, 2006.

Halstead, M., and Walter, K. "Clinical Report: Sport-Related Concussion in Children and Adolescents." *Pediatrics* 126, no. 3 (Sept. 2010).

Hechinger, S. "Another Bite at the Graham Cracker: The Supreme Court's Surprise Revisiting of Juvenile Life Without Parole in *Miller v. Alabama* and *Jackson v. Hobbs*." *Ipsa Loquitur*, online companion to *Georgetown Law Journal*, Sept. 2011.

Henig, R. "Why Are So Many People in Their 20s Taking So Long to Grow Up?" *New York Times*, Aug. 18, 2010.

Hester, R.; Nestor, L.; and Garavan, H. "Impaired Error Awareness and Anterior Cingulate Cortex Hypoactivity in Chronic Cannabis Users." *Neuropsychopharmacology* 34, no. 11 (Oct. 2009).

Hiller-Sturmhofel, S., and Swartzwelder, S. "Alcohol's Effects on the Adolescent Brain: What Can Be Learned from Animal Models." *Alcohol Research and Health* 28, no. 4 (Winter 2004).

Hingson, R.; Hereen, T.; and Winter, M. "Age at Drinking Onset and Alcohol Dependence: Age at Onset, Duration, and Severity." *Archives of Pediatric and Adolescent Medicine* 160, no. 7 (July 2006).

Hirsch, A. "Reflections from the Front Lines: A Career Counselor's View of Emerging Adulthood." *Yellowbrick Journal of Emerging Adulthood* 2, no. 1 (2011).

Hooper, C.; Luciana, M.; Conklin, H.; and Yarger, R. "Adolescents' Performances on the Iowa Gambling Task: Implications for the Development of Decision Making and Ventromedial Prefrontal Cortex." *Developmental Psychology* 40, no. 6 (Nov. 2004).

Humphrey, L. "A Developmental Psycho-Neurobiological Approach to Assessment of Emerging Adults." *Yellowbrick Journal of Emerging Adulthood* 1, no. 1 (2010).

Hyman, S.; Malenka, R.; and Nestler, E. "Neural Mechanisms of Addiction: The Role of Reward-Related Learning and Memory." *Annual Review of Neuroscience* 29 (2006).

Ingalhalikar, M.; Verma, R.; et al. "Sex Differences in the Structural Connectome of the Human Brain." *Proceedings of the National Academy of Sciences* 111, no. 2 (Jan. 14, 2014).

Iniguez, S., and Bolanos-Guzman, C. "Nicotine Exposure During Adolescence Induces a Depression-Like State in Adulthood." *Neuropsychopharmacology* 34, no. 6 (May 2009).

International Center for Media and the Public Agenda. The World UNPLUGGED, 2011, http://theworldunplugged.wordpress.com.

Jabr, F. "Neuroscience of 20-Somethings: 'Emerging Adults' Show Brain Differences." *Scientific American*, Aug. 29, 2012.

Jacobson-Pick, S., and Richter-Levin, G. "Short and Long Term Effects of Juvenile Stressor Exposure on the Expression of $GABA_A$ Receptor Subunits in Rats." *Stress* 15, no. 4 (July 2012).

Jager, J., and Ramsey, N. "Long-Term Consequences of Adolescent Cannabis Exposure on the Development of Cognition, Brain Structure and Function: An Overview of Animal and Human Research." *Current Drug Abuse Research* 1, no. 2 (June 2008).

James, T. "The Age of Majority." *American Journal of Legal History* 4, no. 1 (Jan. 1960).

Janssen, D. *Growing Up Sexually: A World Atlas*, vol. 1, Magnus Hirschfeld Archive for Sexology, last revised Feb. 2006, http://www.sexarchive.info/GESUND/ARCHIV/GUS/INDEXATLAS.HTM.

Johnson, S.; Blum, R.; and Giedd, J. "Adolescent Maturity and the Brain: The Promise and Pitfalls of Neuroscience Research in Adolescent Health Policy." *Journal of Adolescent Health* 45, no. 3 (Sept. 2009).

Johnson, S., and Jones, V. "Adolescent Development and Risk of Injury: Using Developmental Science to Improve Interventions." *Injury Prevention* 17, no. 1 (Feb. 2011).

Jones, R. "Lasting Effects of Endocannabinoids." *Nature Reviews Neuroscience* 4, no. 525 (July 2003).

Kaiser, A.; Halle, S.; Schmitz, S.; and Nitsch, C. "On Sex/Gender Related Similarities and Differences in fMRI Language Research." *Brain Research Reviews* 61, no. 2 (Oct. 2009).

Kaiser Family Foundation, Program for the Study of Media and Health. "Media Multitasking Among American Youth: Prevalence, Predictors and Pairings." Dec. 12, 2006.

Kalish, N. "The Early Bird Gets the Bad Grade." *New York Times*, Jan. 14, 2008.

Kelley, A.; Schochet, T.; and Landry, C. "Risk Taking and Novelty Seeking in Adolescence." *Annals of the New York Academy of Sciences* 1021, no. 1 (June 2004).

Kensinger, E., and Payne, J. "Sleep's Role in the Consolidation of Emotional Episodic Memories." *Current Directions in Psychological Science, Association for Psychological Science* 19, no. 5 (Oct. 2010).

Kerstetter, K., and Kantak, K. "Differential Effects of Self-Administered Cocaine in Adolescent and Adult Rats on Stimulus-Reward Learning." *Psychopharmacology* 194, no. 3 (Oct. 2007).

Killgore, W.; Oki, M.; and Yurgelun-Todd, D. "Sex Specific Developmental Changes in Amygdala Responses to Affective Faces." *Neuroreport* 12, no. 2 (Feb. 12, 2001).

Kim-Cohen, J.; Caspi, A.; Poulton, R.; et al. "Prior Juvenile Diagnoses in Adults with Mental Disorder: Developmental Follow-Back of a Prospective-Longitudinal Cohort." *Archives of General Psychiatry* 60, no. 7 (July 2003).

Knutson, B.; Wimmer, G.; Kuhnen, C.; and Winkielman, P. "Nucleus Accumbens Activation Mediates the Influence of Reward Cues on Financial Risk-Taking." *Neuroreport* 19, no. 5 (Mar. 26, 2008).

Kolb, B., and Whishaw, I. "Brain Plasticity and Behavior." *Annual Review of Psychology* 49 (1998).

Krueger, F.; Moll, J.; Zahn, R.; Heinecke, A.; and Grafman, J. "Event Frequency Modulates the Processing of Daily Life Activities in

Human Medial Prefrontal Cortex." *Cerebral Cortex* 17, no. 10 (Oct. 2007).

Kuhl, P., et al. "Foreign Language Experience in Infancy: Effects of Short-Term Exposure and Social Interaction on Phonetic Learning." *Proceedings of the National Academy of Sciences* 100 (2003).

Kupchik, A. *Judging Juveniles: Prosecuting Adolescents in Adult and Juvenile Courts*. New York: New York University Press, 2006.

"The Life and Death of a Neuron." National Institute of Neurological Disorders and Stroke, last updated Dec. 19, 2013, http://www.ninds.nih.gov/disorders/brain_basics/ninds_neuron.htm.

Lau, J.; Britton, J.; Pine, D.; et al. "Distinct Neural Signatures of Threat Learning in Adolescents and Adults." *Proceedings of the National Academy of Sciences* 108, no. 11 (Mar. 15, 2011).

Lebel, C., and Beaulieu, C. "Longitudinal Development of Human Brain Wiring Continues from Childhood into Adulthood." *Journal of Neuroscience* 31, no. 30 (July 27, 2011).

Lenhart, A.; Purcell, K.; Smith, A.; and Zickuhr, K. "Social Media and Mobile Internet Use Among Teens and Young Adults." *Pew Research Center*, Feb. 3, 2010.

Lenroot, R. K., and Giedd, J. N. "Brain Development in Children and Adolescents: Insights from Anatomical Magnetic Resonance Imaging." *Neuroscience and Biobehavioral Reviews* 30, no. 6 (2006).

———. "Sex Differences in the Adolescent Brain." *Brain and Cognition* 72, no. 1 (Feb. 2010).

Lincoln A.; Caswell, S.; Almquist, J.; Dunn, R.; Norris, J.; and Hinton, R. "Trends in Concussion Incidence in High School Sports: A Prospective 11-Year Study." *American Journal of Sports Medicine* 39, no. 5 (May 2011).

Lise, E. "Girl Brain, Boy Brain?" *Scientific American*, Sept. 8, 2009.

Luciana, M., et al. "The Development of Nonverbal Working Memory and Executive Control Processes in Adolescents." *Child Development* 76, no. 3 (May–June 2005).

Luna, B.; Padmanabhan, A.; and O'Hearn, K. "What Has fMRI Told Us About the Development of Cognitive Control Through Adolescence?" *Brain and Cognition* 72, no. 1 (Feb. 2010).

Lynskey, M.; Agrawal, A.; and Heath, A. "Genetically Informative Research on Adolescent Substance Use: Methods, Findings and Challenges." *Journal of the American Academy of Child and Adolescent Psychiatry* 49, no. 12 (Dec. 2010).

MacDonald, A. "Distinguishing Depression from Normal Adolescent Mood Swings." Harvard Health Publications blog, Harvard Medical School, Sept. 13, 2010, http://www.health.harvard.edu/blog/distinguishing-depression-from-normal-adolescent-mood-swings-20100913335

McCormick, C.; Mathews, I.; et al. "Social Instability Stress in Adolescent Male Rats Alters Hippocampal Neurogenesis and Produces Deficits in Spatial Location Memory in Adulthood." *Hippocampus* 22, no. 6 (June 2012).

McCormick, C.; Mathews, I.; Thomas, C.; and Waters, P. "Investigations of HPA Function and the Enduring Consequences of Stressors in Adolescence in Animal Models." *Brain and Cognition* 72, no. 1 (Feb. 2010).

McCrory, E.; Viding, E.; et al. "Heightened Neural Reactivity to Threat in Child Victims of Family Violence." *Current Biology* 21, no. 23 (Dec. 6, 2011).

McDermott, Terry. *101 Theory Drive: A Neuroscientist's Quest for Memory.* New York: Pantheon Books, 2010.

McQueeny, T.; Schweinsburg, B.; and Tapert, S. "Altered White Matter Integrity in Adolescent Binge Drinkers." *Alcoholism: Clinical and Experimental Research* 33, no. 7 (July 2009).

Meehan W.; d'Hemecourt, P.; Collins, C.; and Comstock, R. "Assessment and Management of Sport-Related Concussions in United States High Schools." *American Journal of Sports Medicine* 39, no. 11 (Nov. 2011).

Meier, M.; Moffitt, T.; et al. "Persistent Cannabis Users Show Neuropsychological Decline from Childhood to Midlife." *Proceedings of the National Academy of Sciences* 109, no. 40 (Oct. 2, 2012).

Mesches, M.; Fleshner, M.; Heman, K.; Rose, G.; et al. "Exposing Rats to a Predator Blocks Primed Burst Potentiation in the Hippocampus in Vitro." *Journal of Neuroscience* 19 (1999).

Mosher, D. "High Wired: Does Addictive Internet Use Restructure the Brain?" *Scientific American*, June 17, 2011.

Mosholder, A., and Willy, M. "Suicidal Adverse Events in Pediatric Randomized, Controlled Clinical Trials of Antidepressant Drugs Are Associated with Active Drug Treatment: A Meta-Analysis." *Journal of Child and Adolescent Psychopharmacology* 16, nos. 1–2 (Feb.–Apr. 2006).

Naveh-Benjamin, M.; Kilb, A.; and Fisher, T. "Concurrent Task Effects on Memory Encoding and Retrieval: Further Support for an Asymmetry." *Memory and Cognition* 34, no. 1 (2006).

Nestler, E., and Malenka, R. "The Addicted Brain." *Scientific American*, Mar. 2004.

Niehaus, J.; Cruz-Bermudez, N.; and Kauer, J. "Plasticity of Addiction: A Mesolimbic Dopamine Short-Circuit?" *American Journal on Addictions* 18, no. 4 (July–Aug. 2009).

Nielsen Market Research. "Mobile Youth Around the World." *Mobile Use Trends and Analysis*, Dec. 2010.

Ophir, E.; Nass, C.; and Wagner, A. "Cognitive Control in Media Multitaskers." *Proceedings of the National Academy of Sciences* 106, no. 37 (Sept. 15, 2009).

Ortiz, C. "Was That My Phone Vibrating?" *Discovery News*, July 10, 2012.

Otallah, S., and Hayden, G. "Concussion in Young Athletes: Heads Up on Diagnosis and Management." Pediatrics Consultant Live (www.pediatricsconsultantlive.com), Apr. 1, 2011.

Pak, T., et al. "Binge-Pattern Alcohol Exposure During Puberty Induces Sexually Dimorphic Changes in Genes Regulating the HPA Axis."

American Journal of Physiology Endocrinology and Metabolism 298, no. 2 (Feb. 2010).

Paus, T., et al. "Structural Maturation of Neural Pathways in Children and Adolescents: In Vivo Study." *Science* 283, no. 5409 (Mar. 19, 1999).

Payne, J.; Stickgold, R.; Swanberg, K.; and Kensinger, E. "Sleep Preferentially Enhances Memory for Emotional Components of Scenes." *Psychological Science* 19 (2008).

Pew Internet and American Life Project. "Trend Data for Teens: Teen Gadget Ownership." Pew Research Center (2009).

Placzek, A.; Zhang, T.; and Dani, J. "Age Dependent Nicotinic Influences over Dopamine Neuron Synaptic Plasticity." *Biochemical Pharmacology* 78, no. 7 (Oct. 1, 2009).

Pyapali, G.; Turner, D.; Wilson, W.; and Swartzwelder, H. "Age and Dose Dependent Effects of Ethanol on the Induction of Hippocampal Long-Term Potentiation." *Alcohol* 19, no. 2 (Oct. 1999).

QEV Analytics. "Survey of American Attitudes on Substance Abuse XVII: Teens." National Center on Addiction and Substance Abuse at Columbia University, Aug. 2012.

Rohrer, D., and Pashler, H. "Concurrent Task Effects on Memory Retrieval." *Psychonomic Bulletin and Review* 10, no. 1 (Mar. 2003).

Rosen, C. "The Myth of Multitasking." *New Atlantis*, Spring 2008.

Rosenzweig, M. "Modification of Brain Circuits Through Experience." In *Neural Plasticity and Memory: From Genes to Brain Imaging*, edited by F. Bermudez-Rattoni. Boca Raton: CRC Press, 2007.

Russo, S.; Dietz, D.; Nestler, E.; et al. "The Addicted Synapse: Mechanisms of Synaptic and Structural Plasticity in Nucleus Accumbens." *Trends in Neuroscience* 33, no. 6 (June 2010).

Samson, K. "Adolescent Marijuana Use May Cause Lasting Cognitive Deficits." *Neurology Today* 10, no. 24 (Dec. 2010).

Savage, R. "The Developing Brain After TBI: Predicting Long-Term Deficits and Services for Children, Adolescents and Young Adults."

International Brain Injury Association, last modified Dec. 6, 2012, http://www.internationalbrain.org/articles/the-developing-brain-after-tbi.

Schochet, T.; Kelley, A.; and Landry C. "Differential Expression of Arc mRNA and Other Plasticity-Related Genes Induced by Nicotine in Adolescent Rat Forebrain." *Neuroscience* 135, no. 1 (2005).

Schoonover, C. *Portraits of the Mind: Visualizing the Brain from Antiquity to the 21st Century*. New York: Abrams Books, 2010.

Schramm, N., et al. "LTP in the Mouse Nucleus Accumbens Is Developmentally Regulated." *Synapse* 45, no. 4 (Sept. 15, 2002).

Schweinsburg, A.; Brown, S.; and Tapert, S. "The Influence of Marijuana Use on Neurocognitive Functioning in Adolescents." *Current Drug Abuse Review* 1, no. 1 (Jan. 2008).

Scott-Taylor, T. "The Implications of Neurological Models of Memory for Learning and Teaching." *Investigations in University Teaching and Learning* 6, no. 1 (Autumn 2010).

Settersten, R., and Ray, B. "What's Going On with Young People Today? The Long and Twisting Path to Adulthood." *Future of Children* 20, no. 1, "Transition to Adulthood" (Spring 2010).

Shapira, N.; Goldsmith, T.; Keck, P.; et al. "Psychiatric Features of Individuals with Problematic Internet Use." *Journal of Affective Disorders* 57, nos. 1–3 (Jan.–Mar. 2000).

Shen, R., and Choong, K. "Different Adaptations in Ventral Tegmental Area Dopamine Neurons in Control and Ethanol Exposed Rats After Methylphenidate Treatment." *Biological Psychiatry* 59, no. 7 (Apr. 1, 2006).

Silva, J. "Constructing Adulthood in an Age of Uncertainty." *American Sociological Review* 77, no. 4 (Aug. 2012).

Small, G., and Vorgan, G. *iBrain*. New York: HarperCollins Publishers, 2008.

Smith, T. "Coming of Age in 21st Century America: Public Attitudes Towards the Importance and Timing of Transitions to Adulthood." National Opinion Research Center, Topical Report 35 (2003).

Snyder, L.; Milici, F.; Slater, M.; Sun, H.; et al. "Effects of Alcohol Advertising Exposure on Drinking Among Youth." *Archives of Pediatrics and Adolescent Medicine* 160, no. 1 (Jan. 2006).

Solowij, N.; Jones, K.; Yucel, M.; et al. "Verbal Learning and Memory in Adolescent Cannabis Users, Alcohol Users and Non-Users." *Psychopharmacology* 216, no. 1 (July 2011).

Sowell, E., et al. "In Vivo Evidence for Post-Adolescent Brain Maturation in Frontal and Striatal Regions." *Nature Neuroscience* 2, no. 10 (1999).

Sowell, E., et al. "Mapping Cortical Change Across the Human Life Span." *Nature Neuroscience* 6, no. 3 (Mar. 2003).

Spear, L. "The Adolescent Brain and Age-Related Behavioral Manifestations." *Neuroscience and Biobehavioral Reviews* 24, no. 4 (June 2000).

Spear, L., and Varlinskaya, E. "Adolescence: Alcohol Sensitivity, Tolerance, and Intake." In *Recent Developments in Alcoholism*. Vol. 17, *Alcohol Problems in Adolescents and Young Adults: Epidemiology, Neurobiology, Prevention, Treatment*, edited by M. Galanter. New York: Springer, 2005.

Spinelli, N., and Jensen, F. "Plasticity: The Mirror of Experience." *Science* 203, no. 4375 (Jan. 1979).

Stamoulis, K. "An Exploration into Adolescent Online Risk-Taking." Dissertation submitted to the Temple University Graduate Board. Philadelphia: Temple University Libraries, 2009.

Steinberg, L. "Cognitive and Affective Development in Adolescence." *Trends in Cognitive Sciences* 9, no. 2 (Feb. 2005).

———. "Risk Taking in Adolescence: New Perspectives from Brain and Behavioral Science." *Current Directions in Psychological Science* 16, no. 2 (Apr. 2007).

———. "A Social Neuroscience Perspective on Adolescent Risk-Taking." *Developmental Review* 28, no. 1 (Mar. 2008).

Stella, N.; Schweitzer, P.; and Piomelli, D. "A Second Endogenous Cannabinoid That Modulates Long-Term Potentiation." *Nature* 388 (Aug. 21, 1997).

Stetler, C. "HBO's 'Girls': A Window into the Psyche of Emerging Adulthood." *Rutgers Today*, June 14, 2012.

Stoolmiller, M.; Wills, T.; Sargent, J.; et al. "Comparing Media and Family Predictors of Alcohol Use: A Cohort Study of U.S. Adolescents." *British Medical Journal* 2, no. 1 (2012).

Strauman, T. J.; Costanzo, P. R.; and Garber, J. *Depression in Adolescent Girls: Science and Prevention*. New York: Guilford Press, 2011.

Sturman, D., and Moghaddam, B. "Reduced Neuronal Inhibition and Coordination of Adolescent Prefrontal Cortex During Motivated Behavior." *Journal of Neuroscience* 31, no. 4 (Jan. 2011).

Taffe, M.; Kotzebue, R.; Crean, R.; and Mandyam, C. "Long-Lasting Reduction in Hippocampal Neurogenesis by Alcohol Consumption in Adolescent Nonhuman Primates." *Proceedings of the National Academy of Sciences* 107, no. 24 (June 1, 2010).

Talavage, T.; Nauman, E.; Leverenz, L.; et al. "Functionally-Detected Cognitive Impairment in High School Football Players Without Clinically-Diagnosed Concussion." *Journal of Neurotrauma* 31, no. 4 (Feb. 15, 2014).

Tamm, L.; Menon, V.; and Reiss, A. "Maturation of Brain Function Associated with Response Inhibition." *Journal of the American Academy of Child and Adolescent Psychiatry* 41, no. 10 (Oct. 2002).

Tapert, S.; Schweinsburg, A.; Medina, K.; et al. "The Influence of Recency of Use on fMRI Response During Spatial Working Memory in Adolescent Marijuana Users." *Journal of Psychoactive Drugs* 42, no. 3 (Sept. 2010).

"Technology Addiction." *From Laptops to LOLcats: Exploring Teen Tech Use*, undated, http://teentechuse.wordpress.com/how-technology-uses-you/hey-how-long-have-you-been-staring-at-this-screen.

Tegner, J.; Compte, A.; and Klingberg, T. "Mechanism for Top-Down Control of Working Memory Capacity." *Proceedings of the National Academy of Sciences* 106, no. 16 (Apr. 21, 2009).

Toga, A.; Thompson, P.; and Sowell, E. "Mapping Brain Maturation." *Trends in Neuroscience* 29, no. 3 (May 2006).

Toledo-Rodriguez, M., and Sandi, C. "Stress During Adolescence Increases Novelty Seeking and Risk-Taking Behavior in Male and Female Rats." *Frontiers in Behavioral Neuroscience* 5, no. 17 (Apr. 7, 2011).

Trafton, A. "Parts of Brain Can Switch Functions: In People Born Blind, Brain Regions That Usually Process Vision Can Tackle Language." MIT News Office, Mar. 1, 2011.

Vaidya, H. "Playstation Thumb." *Lancet* 363, no. 9414 (Mar. 27, 2004).

Vanderschuren, L. J.; Di Ciano, P.; and Everitt, B. J. "Involvement of the Dorsal Striatum in Cue-Controlled Cocaine Seeking." *Journal of Neuroscience* 25, no. 38 (Sept. 21, 2005).

Vassiliadis, A., and Mederich, J. "Digital Withdrawal: I'm a Teenage Tech Addict." *Huffington Post*, Mar. 1, 2012.

Viner, J., and Davae, U. "High Strung and Strung Out: Clinically Relevant Questions Regarding Adult ADHD and Comorbid Bipolar and Substance Abuse Disorder." *Yellowbrick Journal of Emerging Adulthood* 2, no. 1 (2011).

Viner, J., and Tanner, J. "Psychiatric Disorders in Emerging Adulthood." *Yellowbrick Journal of Emerging Adulthood* 1, no. 1 (2010).

Walker C., and McCormick, C. "Development of the Stress Axis: Maternal and Environmental Influences." In *Hormones, Brain, and Behavior*, edited by A. Arnold et al. Amsterdam: Elsevier, 2009.

Walker, Q., and Kuhn, C. "Cocaine Increases Stimulated Dopamine Release More in Periadolescent Than Adult Rats." *Neurotoxicology and Teratology* 30, no. 5 (Sept.–Oct. 2008).

Walsh, D. *Why Do They Act That Way?* New York: Free Press, 2004.

Wargo, E. "Adolescents and Risk: Helping Young People Make Better Choices." ACT for Youth, Department of Human Development, College of Human Ecology, Cornell University, 2007.

Waters, P., and McCormick, C. "Caveats of Chronic Exogenous Corticosterone Treatments in Adolescent Rats and Effects on Anxiety-Like and Depressive Behaviour and HPA Function." *Biology of Mood and Anxiety Disorders* 1, no. 4 (2011).

Weder, N. "Prevalence of Mental Health Disorders in Children and Adolescents Around the Globe." *Journal of the American Academy of Child and Adolescent Psychiatry* 49, no. 10 (Oct. 2010).

Weiser, M.; Lubin, G.; et al. "Cognitive Test Scores in Male Adolescent Cigarette Smokers Compared to Non-Smokers: A Population-Based Study." *Addiction* 105, no. 2 (Feb. 2010).

Weissenborn, R., and Duka, T. "Acute Alcohol Effects on Cognitive Function in Social Drinkers: Their Relationship to Drinking Habits." *Psychopharmacology* 165, no. 3 (Jan. 2003).

Wheeler, A.; Frankland, P.; et al. "Adolescent Cocaine Exposure Causes Enduring Macroscale Changes in Mouse Brain Structure." *Journal of Neuroscience* 33, no. 5 (Jan. 30, 2013).

White, A., and Swartzwelder, H. "Hippocampal Function During Adolescence: A Unique Target of Ethanol Effects." *Annals of the New York Academy of Sciences* 1021, no. 1 (June 2004).

White, A.; Truesdale, M.; Bae, J.; et al. "Differential Effects of Ethanol on Motor Coordination in Adolescent and Adult Rats." *Pharmacology, Biochemistry and Behavior* 73, no. 3 (Oct. 2002).

Wojnar, M., et al. "Sleep Problems and Suicidality in the National Comorbidity Survey Replication." *Journal of Psychiatric Research* 43, no. 5 (Feb. 2009).

Wolfe, J. "Fewer Teens Perceive Risk in Marijuana Use." *Psychiatric News* 46, no. 20 (Oct. 21, 2011).

Wood, J.; Heitmiller, D.; et al., "Morphology of the Ventral Frontal Cortex: Relationship to Femininity and Social Cognition." *Cerebral Cortex* 18, no. 3 (Mar. 2008).

Yeates, K., et al. "Reliable Change in Postconcussive Symptoms and Its Functional Consequences Among Children with Mild Traumatic Brain Injury." *Archives of Pediatrics and Adolescent Medicine* 166, no. 7 (July 1, 2012).

Zhang, T.; Morrisett, R.; et al. "Synergistic Effects of the Peptide Fragment D-NAPVSIPQ on Ethanol Inhibition of Synaptic Plasticity and NMDA Receptors in Rat Hippocampus." *Neuroscience* 134, no. 2 (2005).

Resources

Talking Teenage (www.talkingteenage.com)—Run by two clinical psychologists specializing in work with teens and their parents, this site includes a blog, a discussion board, hot topics, and latest news.

Parenting Teens Online (www.parentingteensonline.com)—This is an excellent information resource for parents of teenagers, offering the latest articles on a variety of topics, including alcohol and drugs, family, health, money, school, and social community.

Sue Scheff Blog (http://suescheffblog.com)—Sue Scheff is an author and parent advocate who founded Parents' Universal Resource Experts (PURE) to help educate parents. She covers a wide variety of subjects from academic cheating to video game addiction.

Annie Fox's Blog (http://blog.anniefox.com)—Reflections on "Tweens, Teens and Parenting" by an author, educator, and entrepreneur.

Travel for Teens and Tweens (http://travel-with-teens.com)—A blog about the best family trips and vacations, including a guide to college visits.

By Parents for Parents (http://byparents-forparents.com)—Includes articles, a blog, a parenting forum, and links to information on everything from weight loss camps and ADHD to bullying and divorce.

Too Smart to Start (www.toosmarttostart.samhsa.gov)—A site developed by the US Department of Health and Human Services, it includes "information for tweens, teens, families and communities to help prevent underage alcohol use."

The Partnership at Drugfree.org (www.drugfree.org)—Guidance, tips, and "stories on the topics of drug prevention, drug abuse, drug intervention, drug treatment and recovery."

National Institute of Mental Health (www.nimh.nih.gov)—The latest information on clinical research and the most reliable resource for education about mental illness and its causes, treatment, and prevention, as well as advice on getting help and coping.

Index

Page numbers in *italics* refer to figures.

About the Authors

Frances E. Jensen, MD, is a professor of neurology and the chair of the neurology department at the Perelman School of Medicine, University of Pennsylvania. She was formerly a professor of neurology at Harvard Medical School, the director of translational neuroscience and the director of epilepsy research at Boston Children's Hospital, and a senior neurologist at Boston Children's Hospital and the Brigham and Women's Hospital. Her laboratory research focuses on injury in the developing brain as well as age-specific therapies for clinical trials development, and she has won many prizes and honors, including Basic Scientist of the Year from the American Epilepsy Society and the Director's Pioneer Award from the National Institutes of Health. Dr. Jensen is the author of more than 130 manuscripts on subjects related to her research and has been continuously funded by the National Institutes of Health since 1987. She is also an advocate for awareness of brain research and has delivered many public lectures and media contributions, including a TEDMED talk; many of these are related to adolescent brain development, its unique strengths and vulnerabilities, and their effects on medical, social, and educational issues unique to teenagers and young adults.

Amy Ellis Nutt is a science writer at the *Washington Post* and author. She is the recipient of the 2011 Pulitzer Prize in feature writing and was a finalist in feature writing in 2009. Her book *Shadows Bright as Glass: The Remarkable Story of One Man's Journey from Brain Trauma to Artistic Triumph* was published by Free Press in 2011.